ARTIFI
INTELLIGENCE BUNDLE: 3
BOOKS IN 1

ARTIFICIAL INTELLIGENCE FOR PEOPLE IN A HURRY

Disclaimer

Table of Contents

14

What is Artificial Intelligence?

The Basic Concept

The basic concept of artificial intelligence is quite easy to understand because it's in the name. Artificial intelligence is intelligence that has been created by hand and not... artificial intelligence is intelligence that is not like real intelligence? Artificial intelligence is a machine attempting to emulate biological intelligence? Actually, the concept of artificial intelligence is quite difficult to nail down to a specific definition.

All right, the basic way that we understand artificial intelligence is an intelligence that is produced by a machine. This has been the standard definition for about a century. The reason why it is difficult to conceptualize a specific definition of artificial intelligence is that of how blurred the line is between biological intelligence and mechanical intelligence.

The Application We Know

The most believable application that we know of when it comes to artificial intelligence is that of the enemies inside of video games. The problem is that this is not really intelligent. Or is it? The way that science defines intelligence is that it is an entity that is capable of attaining new information and then using that information with past information to define current information. Therefore, the primary problem with creating real intelligence is finding a way for machines to learn.

The reason why enemies inside of video games are still called artificial intelligence is the mechanical nature of enemies in video games. The average person who plays a video game will look at a first-person shooter enemy as being intelligent in that the enemy will follow them. Therefore, if the player begins to hide then the enemy will move closer while if the player is out in the open and shooting, the enemy will run for cover after firing a few bullets. This makes it seem like the enemy is actually thinking.

This is the purpose of artificial intelligence; faking the ability to think. The truth of the matter is that enemies inside of video games don't think at all. The easiest way that I can show you how a video game enemy performs these actions is with a little bit of pseudocode.

```
if(enemy.line_trace_end !=
player.collision){
    enemy.position moves closer to
player
}else if(player.line_trace_end ==
enemy.collision){
    enemy.hide * ((Random.new() * 10) +
1)
    if(enemy.hide >= 75){
        enemy.position moves behind
nearest collision
    }}
```

As you can see from the pseudocode above, if the line trace of where the enemy is pointing to does not equate to the player's collision but another object collision, then the enemy will move forward. This brings the enemy closer and creates the illusion that the enemy knows the player is hiding. However, if the player is targeting them and the line trace meets up with the player collision box then the enemy will decide whether to fire at the player or hide behind a collision based on a randomized value. The line trace is the path of motion of the player.

The randomization creates the illusion that the enemy chooses whether they are going to duck underneath cover or they're going to fire at the enemy. However, this is fake intelligence if we consider the fact that there is no past information being involved here. This is current information dealing with current decision-making. Additionally, the machine doesn't actually have any ability to learn about the character that they are going up against. This means that while the machine looks to be intelligent, there is no intelligence at play. Instead, you have an instruction set pre-programmed by a thinking individual that tells the machine what to do given a certain case or a certain circumstance.

22

Now you might think that something like Amazon Echo or Google Home are hubs of real mechanical intelligence. The problem is that these machines are working very similar way to the enemies that we just covered.

Human speech is dictated by rules, you can think of them as the instruction set for how to speak the language. The important thing about rules is that if there are rules, a machine is most likely going to be able to follow those rules. Well, I won't get into the specifics of how recurrent neural networks work, but you can basically assume that the machine is simply analyzing and making probability statistics on what words likely fit together based on those rules. This is a very accurate way of determining which words you, as a person, are going to say whenever you use something like voice typing or when captions are made inside of a video that have been auto translated your words.

Up until this point, we have not had many machines that are capable of learning and using that learned knowledge to further improve. However, we have just now entered the digital age of utilizing something called backpropagation.

Backpropagation is the center point of nearly all machine learning algorithms that are currently at play. A neural network is designed to take database input and perform actions on that information inside of neurons that then produce an output. If the output is wrong, we then change what is occurring inside of the neurons so that we try to get to a more optimized answer. The standard way of doing this is by taking the variables inside of the neural network and creating sets of randomized variables to find the best-randomized variables. This is a very slow process and usually takes a very long time in order to perform. The new method is called backpropagation, what I just described to you is called forward propagation.

Instead of utilizing randomized variables and never knowing if you have reached the ultimate randomize combination, we use calculus and the outputs that the neural network has provided. The information is fed backward into the neural network after the answer is performed incorrectly and the neural network takes the incorrect results as well as the numbers they used to get to those results and creates a combination that is more optimized as a result. Backpropagation represents the first form of machine learning that is actually intelligent. However, the vast majority of neural network applications, like Alexa and Siri, are not normally back propagating networks in most situations. As time progresses, this actually might be changed but backpropagation takes a lot of computing power that most devices simply don't have.

The Applications We're Creating

I've talked about the different types of artificial intelligence; emulated intelligence, probability intelligence, and machine learning. A lot of individuals tend to categorize all of it as artificial intelligence, which is incorrect. If you remember correctly, artificial intelligence is where an entity attempts to emulate the ability to learn without actually learning. Machine learning is clumped in with artificial intelligence, even though there are certain forms of machine learning that actually don't fit in artificial intelligence. You have several different robot projects that are based in machine learning that are no longer artificial intelligence.

However, we are not actually here to discuss that realm of intelligence. Instead, we are talking about applications where artificial intelligence is being used. You have various different industries that are currently massively affected and will be massively affected for the next decade as new artificial intelligence mechanisms come out to affect that industry. You have a self-driving car, robotics that are designed to emulate companionship, and you have loads more that we will talk about in the next chapter. However, we do need to talk about the pinnacle of all artificial intelligence. Artificial intelligence is not possible without information.

When I gave you the example of the pseudocode, you will notice that the choices the enemy actor made were based on information about the current situation. This is true of all artificial intelligence machines because nothing can be decided if nothing is coming in. We, as humans, receive input all the time from the world and artificial intelligence machines need that same input and sometimes even more.

The Information Age

Knowing that artificial intelligence needs information and given the current political climate that this book is being written in, we can conclude that we are in a new era. You have Facebook trying to collect as much data as possible while Europe is trying to stop Facebook from collecting so much information without the permission of the user. These internet-based companies have had unfettered access to the information of our daily lives because it was simply assumed that the user would understand what they were giving up here.

The problem is that the average person is lazy and ignorant most of the time. A person may find that insulting, but that is how the market works. A person buys a washing machine because that's what their parents did, or they don't want to wash their clothes by hand. A person buys a dishwasher because that's what their parents did, or they don't want to wash dishes by hand. There was a man that I conversed with once that did not know where beef inside of a grocery store came from, he simply assumed the grocery store made the beef. I'm not talking about where he didn't know which company shipped the beef to them, I'm talking about he did not know that beef came from cows. This was a person who worked in the government and yet he did not know this very simple, commonly understood fact. There are many people like this. Those individuals who have a lack of what would be normally perceived as commonly understood facts purchasing the items these technology companies make. Markets rely on ignorance and laziness to work. After all, why hire a mechanic if you know how to do it yourself?

People tend to skip the end-user license agreement because it is a huge document that is not very entertaining to read. There is a very small portion of the crowd that will read that document and that portion of the crowd is generally how companies are kept in check. That person will read it, find something horrendously wrong with it, and then share that information with people who don't read it. The general market relies on people not being able to fully understand how things are done. If you bought a t-shirt that was $100 and thought you were getting a great deal, you might find you were instead getting a horrible deal once you know how it was made. You can buy a standard t-shirt off the internet for $5. You can then go take a picture you want, and have it placed on a special paper that you can then heat press onto the shirt for about $5. That means, practically any shirt that you buy that is simply a logo without being any special type of gel and is just a plain image is usually about $10 in cost. Now, there are additional costs, like shipping and branding, but it took $10 to make $100. If you know how to do it yourself, you could have saved an additional $90.

However, you are likely not going to go out and make every t-shirt that you want to own unless you are incredibly stubborn and self-motivated. Most people will simply claim they don't have the time to do it. However, the fact of the matter is that that's a lot of money that is being thrown away due to laziness as many of those same individuals will binge watch a 12-hour season. Restaurants rely on people not making recipes, clothing stores rely on people not making clothes, and tech companies rely on people not creating tech. This is how the market works and how it has worked forever. However, the new commodity is information as information is now how most companies make their

money.

The Fourth Industrial Revolution

Facebook sells your information to advertisers so that

advertisers can better promote the products they are trying to sell.

Google actually does the exact same thing but has different data.

Almost every platform does the same thing, except that most of it has

been allocated to a handful of companies. The standard website will not

go out of their way to create an advertising tree that advertisers can

participate. Why would advertisers go to an unnamed website to

promote their ads? No, most standard websites that are trying to make it

online will utilize Google ads or Facebook ads. This allows the website

to gain money from Google or Facebook by selling advertisements.

Another section of the internet is affiliate links, where people

have audiences that they can direct traffic to specific products. The

most notable of affiliate links are the Amazon affiliate links, which

seem to be practically everywhere. Internet Stars are able to sell products of other people (like an advertisement would) to crowds that would generally be interested in it. This is much more effective than simply trying to blast on the internet that you have a specific product that you want to sell. In this situation, the information is simply that you know what your crowd likes to see and what your crowd is after.

We can't get anywhere without Alexa or Google Voice or Siri because most of our information goes through these three individuals, machines, to produce what we want. We ask Google pertinent questions about our daily lives. They collect information based on our search results and our requests so as to better serve advertisements to us. However, they are not the only customers of information.

You have subscription services like Netflix and Hulu that utilize your watched patterns to determine which videos you are more likely to want to watch. By serving you what you want to watch, these companies keep you on their platform for a longer period of time. In fact, it's so bad that there have been cases where people who watch Netflix go into rehab to overcome their addiction to the service.

Information runs practically all of our businesses as of right now and the person who can collect the most information is often the biggest business on the internet as well as in the real world. If you look at Google, Google is supposed to be a search engine, singular, that provides advertisement on the sides. Yet, they are a massive company being sued for monopoly reasons. If we rewind history back to the 1960s, Google didn't really exist, and you had cable companies and telephone companies being the biggest alongside real estate and other industries. The tech industry, which solely relies on information to actually work, is the biggest industry in the entire world. It's in everybody's home, everybody's kitchen, everybody's room, everybody's car, everybody's road, and I think you get the point.

Our Daily Lives with Artificial Intelligence

Finance Industry

The finance industry is perhaps the most well-known industry for utilizing artificial intelligence before any other industry decided to use it besides the computer industry. It was long thought that stock markets and pricing graphs could generally not be predicted, but a serious investment into prediction algorithms has had sway over the highest companies playing in the stock market.

In fact, it's become so common knowledge that these companies are vying for prediction algorithms that other people, as individuals, are also trying to get their own prediction algorithms. This is because the prediction algorithm can actually be taught to be more accurate than the individual.

However, that's not all the artificial intelligence community has dabbled in with the finance industry because there's quite a bit more. A little bit closer to home is monitoring the habits of the individuals who hold money within a bank. The number one concern when dealing with bank accounts is whether there has been fraudulent activity on those accounts. Due to the fact that banks are often the first financial industries to get hit, besides credit cards, understanding habits that are out of the norm are a key aspect of determining whether a transaction was fraudulent or not. With artificial intelligence, the machine can be trained to understand what is most likely to happen in the case of that specific person having a specific amount of money. For instance, if you have a lot of money at once, you might be the type of person that is a little bit conservative with their money or you might be the person that goes to the mall and blows it all. If you are more conservative on most occasions, the artificial intelligence will get really suspicious if you decide to go blow it all.

The idea of artificial intelligence in the stock market is actually just a segment of what A.I. does in what's known as advisory artificial intelligence. It's a much bigger category and it includes things such as marketing forecasts, audience expansions, and a lot of other business technical terms. The purpose of an advisory artificial intelligence is to train the artificial intelligence to predict what has not happened based on what has already happened. The crux of this type of A.I. is that almost all situations dealing with predicting what has not happened is based on context and so it is difficult to quantify context, but there are several applications of artificial intelligence in advisory roles.

Perhaps the less talked about part of artificial intelligence in finances is the A.I. that helps in finding suspicious patterns with companies. This isn't really talked about because it's not really something that the public is concerned with but rather governmental agencies looking into the public. Previously, the way that white-collar crime could be detected on its own is by looking at the finances of a company and seeing where the money is going. This is a very laborious and man-hour eating task. Artificial intelligence is so efficient and quick that what it could take a hundred-man team to do in a week could be done by the artificial intelligence in maybe a day at the longest. Therefore, A.I. not only is helping out the public with their finances but also helping the legal team in catching criminal activity in companies through their finances.

Sex Industry

Artificial intelligence has made leaps and bounds within the sex industry. If you rewound time to about a decade before this book was written, you would be lucky to meet a chat system artificial intelligence that you could pay by the hour to turn you on. As time has progressed, there have been several sections within the sex industry that have expanded the purpose of artificial intelligence in ways that you would expect and ways that are unexpected.

Of the ways that you would expect, these chat system AIs have gotten better at forming conversations with people. There is a particular type of technology called MMD that when combined with artificial intelligence can simulate a FaceTime interaction. In fact, the aspect of MMD has expanded into another software called Facerig that I see expanding even further into a new type of genre that has recently opened up in Japan. In a product known as the Gatebox, a virtual girl or woman can do practically anything you would want them to do as a virtual companion. You can buy your favorite virtual women or men to put inside of this new technology so that you can come home and interact with them. This is referred to as companionship AI, which is slightly different than assistant AI.

If you had such a technology, you could expect to come home and be greeted with a warm and friendly hello or welcome home from your virtual wife or husband of your preference. This technology would act similarly to Amazon Echo or Google Home in that it would remind you of certain activities or dates, but it would also interact with you on a more personal level. For instance, it would send you text messages saying that it looked forward to you getting home or asking you how you are doing in the middle of your day. Similar to how relationships might work, this artificial intelligence is specifically designed to make you feel better throughout the day.

Carrying this emotional artificial intelligence over into the world of Life-Size dolls otherwise known as sex dolls, we can see that there are several developers looking to imbue sex dolls with emotions. To somebody who hasn't looked at the industry since a documentary was made about it on television, the industry itself has primarily consisted of people who invest money in sex dolls because they prefer a relationship with a doll to be more preferable than a human. One can understand that kind of logic because there are some drawbacks when dealing with a human. A lot of people don't really like interacting with individuals who provide them with negative attitudes and negative opinions, which naturally leads to seeking comfort in other things than the people around them

In the emotional AI of the sex dolls, the human that owns said sex doll can not only interact with the sexual part of the doll but also interact with the emotional elements of the doll. For instance, the doll could be sultry but also caring or it could be a complete tsundere with a little bit of light humor. Essentially, you interact with this sex doll and its personality conforms to What would most likely be best for your personality. Now, there is an option for you to have customization over it for sexual purposes, but if you're looking at it for a companionship role then you can have them adapt to your own personality.

Finally, the most interesting part of this is that there are now devices that allow you to have sex with computers. Now you might think that the previous section that talks about sex dolls referred to having sex with computers, but I am talking about devices that are hooked up to an artificial intelligence to specifically pleasure you. For instance, a recent development of a product for the virtual reality world was called into question in a very weird light. Essentially, the girlfriend of this man had accused the man of cheating because the man used a device meant for masturbation in the virtual reality environment. Imagine a woman being recorded by a man (in the first-person view) masturbating that man and a condom with sensors on it is recording the pressure of the woman's hand. All of this is recorded and when a paying customer purchases this, they can feel the same sensation of that woman's hand on their crotch only it's through a device instead of a human hand. You might be wondering where AI comes into this. Well, there are small segments of the internet where this is not a recording but an MMD character that's using this same device to learn how to properly masturbate the client. That's right, a virtual reality AI can now be used in substitution for, basically, digital prostitution.

Health Care Industry

The healthcare industry benefits significantly from artificial intelligence as artificial intelligence has had significant improvements in image recognition technology. In our everyday lives, image recognition technology provides us with little tags that we can put alongside the faces of the people that we take pictures of. In medicine, the story is a little different as doctors tend to take pictures of bodily organs and parts that might have something wrong with it.

The most common thing that the health industry is using artificial intelligence for is to recognize types of malignant cancer better than human doctors can. The benefit of this is, of course, saving lives, but the truth of the matter is that artificial intelligence does not need to have you be in the same room in order to make a diagnosis. To individuals who do not regularly go to the doctor, you will likely not know why this is important, but this is the reason why Doctor Services are not normally found online. Doctors insist that they get a physical examination from you, which means they often want you in the same space as themselves to give a proper diagnosis.

While it is currently only being used to significantly improve the ability to detect malignant cancers, this technology is going to eventually come down the health industry ladder to the average individual. It is commonly known that an individual simply needs to look into the mirror with their mouth open to see if there are white spots on the back of their throat to give an estimated answer as to whether they have strep throat. Now, doctors often also insist on a saliva swab to ensure that the diagnosis is correct, but many diagnoses can be given in certain circumstances that only require a visual. For instance, a common cold is usually looking at a patient and taking their vitals. However, going to the doctor for such a visit is usually way more expensive than people can justify it. After all, if a doctor visit is 1/3 of your weekly paycheck, you are not likely to go if you make below a certain amount of minimum income.

The problem with this is that people tend to go to work sick even though they should technically be at home. Common experiences from fast food chains has shown that a good portion of the workforce working at minimum wage simply cannot afford to take the time off. Now, obviously, for the sake of not spreading their illness to the food that they are making they should stay home but this is not normally seen as the more important of the two concerns in such a situation by the individual's perspective. To the individual impacted by the minimum wage that is. However, if an individual could go to an online service that was relatively cheap per visit or even just had a monthly subscription that was reasonable, they would probably get seen by an artificial intelligence doctor so that they could get the medicine to make them feel better.

Transportation Industry

Recently, we have actually seen big industries push for automated motors that no one except a computer has to drive. Now, there are a few industries that I want to talk about because the transportation industry is a huge industry, consuming most of the market we know of today. You have consumer cars, the technology going into these cars being equivalent to technology going into a tablet, virtual assistants being inside of cars, and the list really does go on for the automotive industry.

The first one that I want to talk about is the trucking industry. Recently, I got to experience a surface level simulation of what it's like to be a truck driver via the game *Truck Driver Simulator*. I finally understand how difficult it is to drive a semi from the origin to perhaps hundreds if not thousands of miles to the next destination. It involves balancing tons of items that precariously hang outside the back end of a semi. The easiest part of the job is hooking the trailer up to the actual semi. The hardest part of the job is a mixture of corners and dropping the material off at where it needed to be. Psychologically, the hardest part of the job is simply being on the road because some of us travel hundreds of miles every few years visiting different places, so we get a new experience. However, people in these positions often have isolation problems because they can be alone for around a day or two at the minimum.

This does not stop the potential automation for these jobs. This is a decent paying job, usually paying around $20 to $30 an hour as a minimum rate in the United States of America. It's not an easy job, even though it may look easy. As I mentioned before, the hardest part is from the psychologically of driving all of those miles to get to your destination. This job may be automated, in the near future, around this central part; driving to the destination. Elon Musk and Uber have announced they plan to make self-driving a part of their future technology. However, these are the companies that practically everyone knows about. There are several other companies fighting in this self-driving game. Specifically, for semi driving, you have companies who employ these people trying to automate the service.

Now, you have also got to realize that semi driving is extremely dangerous for both the driver and the people around the driver. This is why semi drivers have to go test regularly and generally have one of the hardest driving tests of all the vehicles that there are. However, that doesn't stop these companies from trying to automate an industry that has previously laid untouched by the newest self-driving algorithms.

With self-driving semi's, the overall cost of food items shrinks dramatically. A semi would likely be carrying several items, but let's say that it's 50 items. If a semi driver has to drive for 3 days or 48 hours, then

$$48*30 \big/ 50 = 28.80$$

is added to the products in that semi. Companies don't normally add $28 to a stick of butter, but they will add $46 to a TV. As a semi is the main form of transportation for almost all in-store goods, that cost goes on all products. This means a $1,000 television could be $800 or a bag of cheese goes from $10 to $5.

As I previously mentioned, Uber is in that category as one of the companies that are seriously investing in self-driving cars. As a customer of Uber, one can see that the benefit would be on both the customer and the business side, but not the worker side. Many of the individuals that Uber are either doing so because it is the best option for them or because they're bored in retirement. On the customer side, the ride will get cheaper because you no longer have to pay a driver and on the business side it will get cheaper for the same reason. However, on the business side, you can also profit by being able to have many more cars than you have employed people. The public will almost always want to have a person driving them because of the technophobia and the potential personal conversations one might have while they are being driven somewhere. It's actually something that you kind of pay for that companies don't even realize.

However, if you also recall, I talked about how dangerous automation of driving truly is. Uber has been under some scrutiny because one of their drivers wasn't doing their job properly and the self-driving tester car managed to run someone over (that's how I understand it anyway). This is also there for semi driving because you have an automated vehicle on the road. The only difference is the extremity of how dangerous this is. It is much more likely for a semi-truck to cause an entire roadway of damage than it is a single car. However, as we push forward as a society, it is much more likely to have vehicles with emergency mechanisms so that you can take over should the vehicle not drive properly.

Education

Focus Areas

One of the key problems with education is where the student is actually placing their focus. For instance, I knew a student once that really loved reading but often loathed doing any sort of math. The instructor themselves didn't concern themselves with how this student was failing in this selected area of the study content the instructor was giving. Once I had the student explain to me what he did throughout the day and then also explain to me the feelings he had about the mathematical part of the content, the reason to me as to why he was failing in that part of the curriculum became obvious. He loved reading because he enjoyed going into a world that allowed him to escape, but mathematics is often taught from a listen and do perspective.

Once I suggested that the student not rely on the instructor and instead read the material rather than listen to the instructor, his mathematical scores significantly increased. The reason for this failure was simply due to how it was delivered. I can foresee an artificial intelligence system that takes over consultancy as we have actually seen in artificial intelligence machines built to diagnose patients. This field of artificial intelligence consultancy is relatively new and still experimental, but I can find something as simple as what I found then an artificial machine, with the same variables, will undoubtedly be able to draw a similar conclusion. However, the conclusion will be brought about in a significantly different way until machines can understand Context. Context is the last field of abstraction left in the artificial intelligence pipeline for improvement.

Teaching without Teachers

This leads us into the other area in which artificial intelligence is both currently affecting and will affect teachers in the future. As of right now, artificial intelligence is being used to predict which areas of statistical analysis are either the weakest or greatest. Therefore, if a lot of students seem to be failing at a very specific part of your mathematical class, say algebraic functions, this would only be

48

noticeable to a machine that could calculate the number of students multiplied by each test answer. Now, that may seem like a very simple Equation to perform, but the equation is much more complex than that. You would need an algorithm that can quantify each answer as both parts of a group and an individual, which would lead into determining whether a group of errors is a common pattern or a group of independent errors is a pattern. Such a distinction is easily made between two students.

For instance, if a student of Class A gets functions wrong and students of class B manages to get one function wrong, we would naturally conclude that students of Class A is having problems with functions, but students of Class B simply got the answer wrong. By having artificial intelligence be able to look at this, artificial intelligence could then create a course plan for students individually based on the topics the artificial intelligence is teaching. The purpose of a human being in a teaching position instead of mandatory reading is so that there is a guide for the person that is learning and a person to set the goals of the class. An instructor is a person who simply tells the class what they need to do, a teacher is a position held by those who act as instructors and those who guide students through those instructions. However, having an artificial intelligence that could figure out whether the class as a whole is having a problem, an individual is having a problem, or a particular topic is doing rather well you could have that same artificial intelligence change its own curriculum plan to better teach students. The artificial intelligence can already act as an instructor by default, as many help prompts have told us in the past, but by being able to optimize a curriculum the artificial intelligence also becomes something that can guide those instructions.

Journalism 425

Aggregate Information

One of the main purposes of Journalism is to aggregate information into a central location and generally provide it objectively. Notice that I said generally because in the current journalism environment, this is rarely done. It is rare to see journalists simply presenting the facts and then distinguishing between the facts and their opinions. The best example of aggregate information being carried out by artificial intelligence is the Google you use every day or most of you. Google does something called sending out a web crawler, which is a fancy way of saying that an advanced algorithm designed to find websites on the internet and include them inside their search listings. Over time, they have created algorithms to determine the accuracy and importance of them, but the primary example here is to show that artificial intelligence has been aggregating information in the past, but this is known as dumb artificial intelligence. New artificial intelligence will be able to collect specific information, most notably the information related to news. It will then be able to follow leads of that news on the internet to see if it's related to any sort of data. For instance, statistics on crimes or number of bakeries in a city on average. This leads us into the current climate of Journalism as we know it in the political sphere.

Accuracy Judgement

There has been a massive push to censor what's known as fake news, but this has a few problems attached to it. The big problem that everyone practically knows about is how to determine what is fake and what is not fake. The obvious answer that most people reach for is that if it is true, more people will be talking about it. However, you have flat earth believers and round earth believers that have constant debates about Earth. If you were to go and search the amount of Flat Earth information, there are more websites that consider Flat Earth to be a reality and there are practically none about round earth because those who believe it is round don't feel the need to convince people. The round earth believers simply assume that anyone with a higher IQ would reasonably believe this and just accept it as fact. Therefore, in the online space, you have more people talking about Flat Earth being a fact than you do round Earth people. Following this obvious law, you will quite literally censor the people who believe in round Earth. Now if you felt one way or the other about the information I just provided beyond completely static, you will understand why it is important not to censor ideas. This is actually the reason why the free speech in America is so protected because the public generally wants what they consider to be crazy out where they can fight the idea rather than have that idea in a

52

position of power and not be able to stop it. There are artificial intelligence systems being created to determine fake news, but the specifics of those facets of artificial intelligence haven't really come to light just yet.

Automated Social Media

While the specifics of accurately judging journalism aren't known, artificial intelligence can easily take over automated jobs. For instance, when sharing an article amongst the many social media websites, most websites simply have a single button attached to the publish button that allows them to just instantly post to all of the websites at the same time. This makes the whole process easier and thus saves the need for controlling whether websites get published on social media on a regular basis.

However, what has not yet been achieved but has mostly been theorized is automation of social media activity on those social media websites. The difference between sharing a post that was written and being active on a social media site is the difference between writing a story and writing a tweet or a Facebook post or whatever website you use to update everyone in your life. Artificial intelligence is currently at a crux in existence because we don't quite yet know how to define the context. That does not mean that a social media account wouldn't be able to regularly post tweets given a set of rules.

For instance, one of the common rules that I was taught during low-level grades is to take the question being asked and reword it as my first sentence. This is a rule that can easily be followed by an artificial intelligence as we have many reword websites that allow you to take sentences and reword them. If someone posts on social media, an artificial intelligence algorithm would be able to take what they said and reword it according to rules. Additionally, they could post things that your website might talk about in the next day based on titles that you wrote today. Essentially, the artificial intelligence would be generating leads that would entice users to follow that account because they're interested in what you said. Social media for journalism is both a medium for spreading information but it's also a medium for advertising the journalism that the journalists are doing. This is already being done today in marketing campaigns by activist groups to hide how small that group might be. We heard a lot of rhetoric in the presidential election of Donald Trump about social media bots doing just that.

Agriculture

Better Forecast Predictions

One of the most wasteful things that a farmer has to deal with is just how much they need to plant as well as just how much they need to harvest as well as when they harvest. In the United States of America, a lot of the food waste goes to food that is neither appealing or used. In fact, this is a common problem in developed nations in the world because most of the developing nations work on a monetary system. If you harvest too much, you will end up wasting what would essentially be the monetary value when you can't sell it. If you don't plan enough, you might lose out on potential gains you could get had you grown enough.

This is an equal problem for the people eating that food because even though you only need to produce a certain amount of food so that the market sustains itself, there are people in developed nations still dying from starvation. It is an odd problem when you have so much food that you are throwing it away and yet you continue to allow starving people to die. The problem is that the market isn't entirely predictable and so you will have waste, or you will have shortages, but you will never break even. Artificial intelligence may not be able to break even, but it can provide farmers and those in the agriculture business with better predictions on just how much they need to produce. The easiest way that many of them use is simply to grow as much as they can and then use what they can't sell. This works out for a lot of small farmers, farmers with around 100 acres to 500 Acres. In these small localized areas, they are still able to use most of what they grow up and make a decent living off of what they sell. The problem really comes into play when you go above 500 Acres. A 4-person family is not going to be able to consume 900 acres of corn in the next 6 months unless they eat nothing but corn for the next 6 months, but if they sell another type of crop it is virtually impossible. Also, that family will probably die because corn doesn't have much of a nutritional value. In these circumstances, it would be better to figure out a trend line for what they need to plant and how much they need to harvest as well as when to harvest. While farmers do a fantastic job at providing food for a nation, not a lot of them are statisticians by trade. In fact, because of how much professional statisticians make in various roles, it would be more lucrative to be a statistician if you were just out for the money. Therefore, the best alternative is to provide a product that will make those predictions for you and this is where artificial intelligence comes up. Artificial intelligence already has a huge involvement in the stock market, so utilizing it for something a little bit more predictable like farming has been an easy crossover.

Advanced Growing Techniques

Farming doesn't really have many scientific advancements as to how food should be grown but if the farming is done in a controlled environment, these variables can be handed off to an artificial intelligence machine. The reason why such a fact is important is due to how artificial intelligence machines are designed to optimize whatever they're handed. A lot of farmers work is actually based on guesstimation or estimated guessing. For instance, many farmers will look for a specific temperature that holds consistency to plant items that are temperature sensitive. You can only grow potatoes during certain seasons based on your location. An artificial intelligence machine could draw on forecast weather and predict weather patterns, which would lead to faster planting of the productive plant.

Because artificial intelligence can be handed Sensors, they can also handle when the soil needs to be given a specific type of nutrient. Normally, farmers rely on the look and condition of a plant to determine when that plant needs more of a specific element. Usually, the plant will show signs of yellowing and wilting. However, if you have sensors in the soil and you have visual open recognition monitors looking at plants, you can teach artificial intelligence to look out for the same things. A crop may actually go for a couple of weeks without the proper nutrition because the farmer simply doesn't notice the signs that quickly, but an artificial intelligence machine would be able to know almost immediately that something has gone wrong and what has gone wrong on the day that it begins. This would allow Farmers to produce more product and reduce the amount of waste, which would generally benefit everyone. This is why, within the past decade, this new form of artificial intelligence has taken root in the more tech-savvy farmers in the market.

Enhanced GMOs

The last bit in which artificial intelligence can help is when developing new genetically modified organisms to better grow for their environment. There's been a lot of stigma around GMO's, but the truth of the matter is that a lot of what the world eats today is made up of GMOs. For instance, most corn could not be sold on the scale that it is without being a GMO of the original corn plant and a weed. In fact, some consider corn not to be a vegetable but a weed because of its origin of being mixed with a weed.

We partake in genetically modified organisms everywhere even though it's considered bad by some. Our children are genetically modified because they are neither all of us or the significant other required to make one. Our food has been optimized to grow more and grow bigger by mixing it with breeds that last longer, provide more food, and provide bigger food. It wasn't until recently that we began using a lab to create GMO products, but we have been doing GMO practices for centuries at this point.

Knowing this and knowing that it is a science, we can build artificial intelligence machines that predict better combinations for the future. This leads to more food and thus more product, which also means that we can sustain more population as a result. This is actually already being done, it's just not widespread yet and a few massive corporations have dedicated their resources to perfecting this.

Law

Better Defenses

There is currently an artificial intelligence known as Lisa, which acts as a robotic lawyer for creating legally binding contracts. Due to artificial intelligence being able to access a wealth of information, they are able to formally design rules based off of past rules. It's actually not that difficult for an artificial intelligence to create a binding contract because of a few elements.

The first element is that many contracts are almost identical, and you can see this when you deal with websites that have autocomplete forms with the contract. The second element is understanding the parties involved and the legal relation of the importance in a document. The last element is because, ultimately, most parties will proofread any document provided by a robot lawyer.

The first part is that many legal documents are identical, and you can actually see this commonality when it comes to software. There have been a few notable software companies that have just stolen the contracts developed by bigger companies as the way to save on money. They need, generally, the same things because they are a software business and so most of the elements affecting other software in legal terms should technically affect their company. Due to the standardization of legal documents, it is easy to figure out what contract you need. This choice can also be easily made by AI based on questions, such as "What type of company are you?" or "What are you attempting to achieve with this contract?".

The second part is that the relationships between parties are generally easy to understand. A software usually has one type of relationship to a user just as a real estate manager has one type of contract with the person that hired them. There are more complex relationships, but for most of everybody, there's usually a one-to-one relationship when it comes to a contract. This person wants something from this other person and needs a contract for it. Due to the simplicity, Lisa is currently able to provide non-disclosure agreements and property contracts as a result of this.

The final element is that we simply don't trust robots completely yet because we've been given plenty of the examples not to do so. Therefore, all contracts made by an artificial intelligence will be proofread by the person who asked for it and then the person who asked for it will provide feedback. This allows the company that develops this artificial intelligence to further optimize their artificial intelligence to better fit the needs of their customers. Therefore, as customers come in the artificial intelligence not only gathers data from those customers but also gathers data from customers with complaints.

Better Rule Enforcement

It is generally seen as bad practice to be a bad lawyer for criminals, but it also is seen as a bad practice to be a good lawyer for criminals. You might think that this sentence is conflicting, but I am talking about two different perspectives. To other lawyers, it is a bad idea to be a bad lawyer for criminals because then you are just a lawyer bad at their job. To the public that doesn't like the criminal, you are a lawyer defending a criminal and so you are often seen as bad as the criminal and sometimes even worse.

Robots don't have to care about this perspective issue because they are robots. This allows them to truly be impartial when putting together a case to be used in court. There is an application known as the *DoNotPay* application, currently found on the Apple Store, that allows you to get free legal advice. There is a legal firm known as ROSS

Intelligence that allows legal teams to analyze documents for legal processing much faster than they would be by a team of humans. These are new technologies that are currently experimental but have shown not only can they do it faster and better than most humans, they can also do it for practically free. That isn't to say that companies won't try to make money off of these technologies, but most of what I just introduced is actually charged to the customer. Normally, a law firm will charge after the first hour for consulting. A law firm using humans to review documents will usually bill a customer for that. By having robots do this, not only do the law firms gain unbiased reviews to the information but customers also don't have to pay exorbitantly high amounts of money to hire a lawyer. A small firm could use these technologies to help their local city have effective defenses to potentially larger companies or people.

Business Processes with Artificial Intelligence

Smart Virtual Assistants

We have had virtual assistants for almost a decade but almost everyone agrees that virtual assistants aren't very useful. The truth of the matter is that they are not really built to be that useful. However, it depends on what you want out of a virtual assistant as to how useful they can be. You see, you can utilize virtual assistants to schedule meetings, read emails, and generally do a lot of office related items with.

The reason why most people don't find virtual assistants very useful is that they often just utilize it to answer questions they might have. While that is definitely useful, a virtual assistant can take notes on business meetings, act as a recorder for sessions, send mass emails to everyone on the business account, and do much more just as a real assistant would. The only problem is no one really knows the best way to talk to a virtual assistant. Most virtual assistants are activated by speech, which is a contentious issue whenever you have any type of accented language involved. The way that these virtual assistants are trained is by having a bunch of voice actors come in and use their voice to match up words to sound patterns. You can actually train a virtual assistant to recognize your accent, but it takes a lot of work. However, you will find them generally used for doing any office related activity that you might have in your business. You can even have your virtual assistant read a book to you if you really wanted to, it really just depends on what you want out of it and most people find virtual assistants to be almost as usual as a human assistant. The only thing that I can think of that a virtual assistant can't do for you just yet is write an email.

Market Research

A huge portion of machine learning, and artificial intelligence is specifically designed to automate market research and refine it. Most market research done by artificial intelligence, provided that it has been trained correctly, is more accurate than the Market analyst can be at making the same prediction. This is because the computer analyzes more numbers faster than the analyst can and the computer doesn't make mathematical mistakes. Normally, when you ask a company to give you a market report they will come back in about a week with a long piece of literature to explain the market for you. However, if you utilize an artificial intelligence to do the same thing, you will often find that the report takes a few minutes to generate provided you have the right data.

Additionally, thanks to algorithms like the K cluster algorithm, you can find General Trends and successes where it would normally seem like nothing was going on. This is often used to identify successful companies in areas of interest that would normally go unnoticed.

Chatbots

Perhaps the best source of artificial intelligence inside of the business environment comes in the form of chatbots. If you want to cut down on the amount of Human Resources allocated to customer support, then a chatbot is usually the way you want to go. Chatbots that are artificially intelligent and not just pre-programmed lines of script have a wonderful way of convincing other humans that they are talking to a human. Most notably, the response time from a chatbot versus a human is significantly lower and thus many more people come out of talking with a chatbot feeling a lot better about the company.

Having said that, chatbots usually aren't that intelligent. If you are going to hand chatbot something to do, you should probably hand it the most common items that your company gets called for. The reason why you want to keep it simplistic and why you want to find a way to remove the most common items from the call list is because chatbots are not intelligent but will reduce the amount of calls your call center will get.

Email Autoresponders

As I mentioned earlier, virtual assistants are currently known for being able to provide the basic necessity of sending out emails. However, you likely thought of those remailers that simply sent out emails responding to password or email resets. The autoresponders currently coming out are a little bit more advanced than that. You have autoresponders capable of opening help desk tickets, answering common questions like software pricing, and even ones that will inform clients of available times they can make appointments. Some are so good that an actual person rarely needs to be in contact with a customer, which saves countless hours on the customer service end.

While I say this, we also know how horribly wrong such a service can be. In truth, this is a common occurrence as artificial intelligence can't rethink an existing solution. Therefore, when an oddity occurs the artificial intelligence cannot self-correct, which can increase the dissatisfaction a customer can experience. This is often remedied by ensuring that the customer only interacts with the autoresponder for a given number of messages.

Taxes

Taxes are the bane of everyone except the oddities that find it fun, but everyone finds something fun, right? Artificial intelligence exists based on rules, which means that a tax system, an exorbitant system of rules, is kind of like the bread and butter of A.I. However, there is a slight crux to Tax A.I. Due to the human nature of constantly changing practically everything, taxes change on a yearly basis, which has to be accounted for by the software. This has been what has taken software companies so long to create an efficient A.I. that's capable of doing an individual's taxes. Therefore, this is one of the new frontiers for A.I. as companies have begun rolling out new software to fit this need.

However, depending on how it is deployed and implemented, a company that uses the software may not even need to fill out paperwork. Since you have to give the A.I. your input, it could be capable of doing all the taxes for you eventually. However, most tax A.I. is centered around determining how taxes will affect the business of the one forecasting the prediction and/or the economy that individual/company is in.

Self-Driving Cars

What are Autonomous Vehicles?

What it means to be Autonomous

To break down with a self-driving car is, we first have to actually understand what it means to be autonomous. The actual word itself is comprised of two Greek phrases. The first phrase, *auto*, means self. The second phrase, *nomous*, means customary or law. Therefore, one can naturally conclude that the phrase literally means self-law or self-custom. By itself, it doesn't have much meaning. It is only when applied to an entity that is capable of thinking that autonomy actually has a purpose.

As a person, you likely have a job or you likely have responsibilities that you perform by yourself. You are autonomous in cases where you decide what you need to do. Autonomy refers to the ability to assign laws to yourself so that you follow them. This means that whenever you do a specific task, how you go about that task is a practice of assigning yourself laws.

Therefore, an autonomous vehicle refers to a vehicle that is capable of assigning itself laws without the need for human interference. For instance, if you were driving a vehicle that had no break, would you hit one person, or would you hit 10 people? This is a common morality question and it rightly has to deal with vehicles and whether it's a good idea for autonomy. Most people believe that if left to choose on its own, the vehicle would simply choose the single individual because of quantity. To those who see robots taking over in the future, obviously, the choice is 10 people. However, the machine, knowing that it might hit a person could choose an option that probably wasn't relevant to you at the time of me introducing this morality question. The machine would simply crash on purpose, which might damage the driver but is not likely going to kill them.

The morality question usually involves trains and not being able to make such an additional option in such a case. However, in the real world, that question is rarely going to actually exist. The ability to see an additional option and be able to take action on it without us assuming the choices it will have allows a vehicle to be autonomous. We, as humans, would commonly assume that there are only two options, but the vehicle would see additional options. Therefore, if the choice was left up to us, lives would be lost unnecessarily. This is the greatest point of autonomy for machines, allowing them to choose the best option without humans choosing that option for them.

The Variables in Vehicles

However, choices are based on variables and if we have an autonomous vehicle, that autonomous vehicle has to have variables. The first variable is the road itself, which is to say is it a straight road or is it an intersection or is it curved? All three types of Roads would lead to a different reaction from humans, which means that the shape of the road is ultimately the first variable that this autonomous vehicle has to deal with.

Just as it has to deal with the road, it also has to deal with the driving laws of the state or country or province that the car is in, which includes the speeding limit. This is so that the autonomous car follows human-made laws that have ensured human safety, which is the point of training a vehicle on the road. An autonomous vehicle will always be able to drive straight, it's literally just one variable that's being affected. It's only when we introduce the need for safety that extra variables are added it.

Therefore, along with knowing the speed limit, the autonomous vehicle now needs to be able to see. The vehicle needs to be able to see so that humans walking in front of the car will cause the car to stop. However, humans don't really see in the same way that computers see. There is a fantastic library called Open Vision that allows a computer to recognize elements that are important in the visual realm. The way that the computer would recognize that there are humans is if there is a bipedal shape crossing its vision. This would allow the car to react within the time that it took to recognize the bipedal shape.

The last variable that vehicles need in order to be autonomous is the weather, but this can actually be done with the internet, right? Actually, not really if you want it to be fully autonomous. There are places out in the world that don't have access to the internet, even in countries where internet seems to be practically everywhere. In such cases, you don't want an autonomous vehicle driving through the rain if it doesn't know that it's raining. Humans react differently in the rain because they need to be more careful as their vision is limited, which, by the way, the computer will have almost the same amount of limited vision with a few exceptions. For instance, humans cannot see in infrared whereas computers simply have a lens switch. Humans cannot have night vision, which is something that cameras can have. This allows the Open Vision system to see in environments that would normally be difficult for humans.

A Combination of Brilliance

Therefore, all in all, in order to create an autonomous vehicle, you have to have a combination of extremely brilliant inventions. The vehicle needs to be spatially aware so that means that proximity sensors need to be inside of the vehicle. The vehicle needs to be able to see humanoid shapes as it is driving along so as to not kill anyone. The vehicle needs to be able to understand the weather that it's driving in so that it can drive appropriately, which means that it needs to be able to have weather forecasting built-in. These are separate industries coming together to make it possible for a car to drive itself, which is one of the biggest feats of science ever to be created.

Common Fears

Replacement of Transport Drivers

The primary fear when it comes to autonomy in vehicles is the replacement of jobs. The problem with making advances in science and society, in general, is that there is always going to be a loser as a result. As of right now, a lot of people are worried that some who work as Uber drivers would be replaced by these machines since then Uber wouldn't have to pay for humans to drive around. This argument is similar to when Uber came out and taxi drivers were worried about being replaced.

A lot of people don't realize that a lot of people are very suspicious, which means that this new technology is not going to be widely accepted. It's going to have a few decades before any of us even trust it as much as we do the Windows operating system. Keep in mind, many of us still don't trust Windows operating system. Therefore, no, it's not going to replace Transport drivers, but it will supplement the availability of Transport drivers. I, for one, wouldn't mind being able to drive to locations for literal pennies. At least I know that an autonomous car is not going to mug me or rob me, which is not to say that the Uber driver will, but the likelihood is zero with the autonomous vehicle and unknown with the Uber driver. It would provide individuals who use Uber as a cheap option to go around town with an even cheaper option and thus the Uber driver would then become a premium experience. Therefore, when autonomous vehicles come out, you would actually see a rise in Uber payments because the human element would be appreciated more by the customers who want that human element.

Unblamable Death

The secondary fear when it comes to autonomy and vehicles is who is going to be blamed for the death. As of right now, there have been a few companies that have made accidents that have resulted in the loss of lives and injuries in other cases. In those cases, the company that put the vehicle on the road has been sued and thus that is the end of the tale. The truth of the matter is that this is what would happen with autonomous vehicles.

The problem that seems to be bothering people is that if someone owns a vehicle that is autonomous by themselves that the person is responsible. However, if the vehicle is meant to be driven by itself then it is the manufacturer who is at fault because they are the one who accidentally left something out that resulted in the incident. Now, there is the highly likelihood that the company will shove the responsibility to the user as the user is supposed to pay attention and ensure that the vehicle is driving, but this would also make the vehicle pointless in many people's eyes. When people think of autonomous vehicles, they often depict what is shown in movies where the humans don't even bother learning how to drive. They simply call up their vehicle and go to another location.

However, companies do not want to assume risk in their vehicles and will likely treat autonomy the same way that companies treated cruise control. Cruise control allows you to maintain a certain speed, but you are still responsible for what occurs in and outside of that car as it is related to that car. For autonomy, companies will likely sell it as a feature much like cruise control that you will still be responsible for keeping an eye on the machine. This ensures that the driver of that vehicle is still responsible for any death that vehicle causes because it is a feature, not something to be relied on 100% of the time. It's a very clever way to get out of responsibility for when software doesn't work on the company's part.

Google Map-esk Incidents

The last fear, which is the less popular fear, is that the vehicle will act very similar to incidents that have happened in the past with mapping Services. I mentioned Google Map as it is one of the most popular methods to navigate around cities and countries. However, there have been moments where Google Maps is not entirely accurate. For instance, if you have a private community, Google Maps is not allowed to actually take photos of what is inside of that community. A lot of people who suddenly join a private community often find themselves having to give delivery drivers directions on how to navigate a neighborhood because Google Maps is not allowed in there.

However, Google Maps can actually lead to the wrong directions. For instance, Pokemon Go had an incident occur several times with its application because people simply wouldn't pay attention to the outside world as they did their exercise. This led to the injury and death of many people because the maps themselves were flat and didn't show that there was anything to worry about. This was, rightly so, the fault of the person who didn't look up from their phone, but the people still were outraged that this occurred. The people blame the company, when it was really the stupidity of the individual who didn't look up. This is the only fear that doesn't have an immediate solution to it and it's only because people get outraged when they are told they have to be responsible for it.

Benefits of Autonomous Vehicles

No More Drunk Driving

The benefit of having an autonomous vehicle is that you no longer rely on the human inside of it. Numerous deaths and accidents are caused by the humans inside of the vehicle not being fully aware. While the person inside of the vehicle might be drunk an autonomous vehicle would be able to navigate that person home by itself without putting anyone at risk at the same level of the individual who is drunk. This means that there would be a significant decrease in deaths and accidents caused by drunk driving.

Optimizing Traffic Incidents

There are several traffic incidents that are often caused by humans not being aware of their surroundings. For instance, a person can only look so far outside a vehicle and many companies stupidly like to put things in front of their business that make it more difficult to drive. Instead of having the view from the front of the vehicle, which would be the safest way to view turning left or turning right, the human is usually in the center of the car. Small incidents like this could be avoidable with an autonomous vehicle because every inch of the car could then be loaded with sensors and cameras that would allow it to react to its environment.

Software Knows All Laws

Furthering that point, a lot of accidents and tickets as well as jail time is caused from an ignorance in the law that the human is supposed to follow. For instance, everyone knows that jaywalking is technically illegal, yet it is rarely enforced. In fact, if it wasn't for jaywalking, cars that were being built for autonomy would not normally need to handle cases where they might hit a human.

Cheaper Taxi/Government Taxi

I talked a lot about Uber and how the autonomous version of Uber would result in a cheaper tier of Uber drivers. However, that could also mean that the government could, instead of providing a bus for localized transportation in Big City, they could instead provide car driven Uber-like drivers. This would be much more helpful to individuals who need to get to work on time as they would be able to rely on the government service that drove them to work. This would affect a lot of people across the board, but it might also have some implications as a result.

The Significantly Disabled Can Own Cars

In the cases where a human is blind or in the case where a human is deaf or in the case where there is something physically wrong with the arms or limbs associated with driving, this vehicle could do it for them. A lot of the individuals, if you are disabled, usually have the right to drive stripped from them and this makes their life significantly harder if they need to go significant distances. Autonomous vehicles could, essentially, provide those disabled individuals with a way of owning a vehicle and normalizing their life after whatever they've experienced.

How will it Impact traffic?

No More Traffic Jams

This one will really only apply if almost all the vehicles on the road are autonomous vehicles. This is something that can be seen in many sci-fi films and novels where cars are able to control how the traffic flows. Given sufficient data and perhaps a network connection to the motor grid, you could technically create a system where no traffic ever has to stop. This is because the only reason why traffic stops is often due to the fact that humans need time to let other humans go the way they need to go. However, it is possible for cars to create a situation where no cars need to ever stop. It would be different than the way we experience traffic now because it would fully rely on automation. However, in the beginning steps of this automation, we would likely see a lot less traffic jams that were caused by accidents or incidents where the police pulled an individual over.

Less Death

As I had mentioned with the drunk driving and the ability to be a driving law book, there would be a lot less death due to traffic problems. This is primarily due to the fact that humans react slower than machines and so lives would be saved faster in emergency situations. Additionally, things like accidentally going before the light actually turns green would grind to a halt for most cars.

Artificial Intelligence and the Job Market

How many jobs will be replaced and why you should care?

All Jobs Will Be Replaced

Let's cut to the chase because every time that this topic comes up, it's really associated with the job that's being replaced. The problem with this topic is that people see the short-term result of what a specific technology will do. The truth of the matter is that all of the jobs in current existence will be replaced, given enough time. It is extremely easy to define and put into place machines for cooking hamburgers. The fact that we refrigerate hamburgers that are pre-made proves this point. Machines can already do most of the basic work that we need them to do.

It is not difficult to automate something, but what is difficult is choosing what needs to be automated. If you have ever talked to an accountant before, you will find that most businesses have different situations for their financial needs. It may be very easy to replace the hamburger maker or the cook in the kitchen with a robot, but it is much more difficult to replace the person who can assess the situation. The problem with artificial intelligence is that it doesn't have something that allows it to understand context.

Whenever you go to pay your bills, you may not pay your bills on time on purpose. In an automated system, the bills will be paid on time every time. However, you may need to wait a week to pay a specific bill because of a certain reason. Tons of people delay paying bills because of reasons and so one problem that robots have is understanding context, the reason why something is being done. This means that even though all jobs will eventually be replaced, the jobs that will be replaced last are the jobs that require context. You cannot automate the process of building a full-scale website, you can automate the design process, the building blocks, and many of the different elements of a full-scale website but that website changes based on the company needs.

Even the Creative Jobs

This means that eventually even the creative jobs will be replaced once artificial intelligence machines can understand context. Here's the problem though; why does it matter? Why does it matter that jobs will be replaced? Jobs exist so as to continue our survival and basically to give us something to do until we die. It's not really a bad thing if all the mandatory jobs are replaced by robots because there is always going to be something else to do. Okay so you don't have to make hamburgers, but you can choose to create a shop of human-made food. That will become a specialty, shops that pride themselves on using only Human Services. Sure, you could automate a car fully, but a car is not going to speed down a runway at top speed to give you a drill thrill. Robots are designed to repeat repetitive tasks, things that you do for fun are things that only humans can do will exist. Only humans can make human made food or human made clothing, something that will be seen as the new fashionista style. The job market is always going to exist and there will always be something to do, you just have to have the right perspective.

What do you need to know to implement A.I.?

Easier Than Ever

The first thing that you need to know and understand is that you are standing on the shoulders of giants. A lot of people don't like to start out with this because they might think it's a little arrogant, but if you are just getting into artificial intelligence then you need to understand this. There has been a lot of work done in the past two decades regarding artificial intelligence, which means you are going to need to do a lot to get to where the frontier is at. That isn't to say that you can't do it within a reasonable time, it's just that you need to appreciate the complexity of this industry. Additionally, you need to understand that it has taken a lot of work to make things easy and while there are very easily implementable tools out on the market, understanding the core mechanics of how neural networks work is key to using these tools. The tools simply allow an individual to get the work they need done without having to deal with much hassle, understanding how those tools work is still something you're going to need.

Algebra to Calculus

The second thing that you're going to need to know is a variety of mathematical skills depending on what you want your artificial intelligence to do. If you want your artificial intelligence to simply forecast the next week's stock prices, you'll mostly need to know statistics as well as maybe a little calculus. If, on the other hand, you want to utilize artificial intelligence to generate 3D pieces of Art then you may need some geospatial mathematics along with a little bit of discrete mathematics. There is a wide range of mathematical skills that may be required depending on what you want to do with it, but the reason why I specifically state algebra to calculus is because you will at least need to know algebra. Neural networks are designed with the understanding of the slope-intercept form as the most basic form of a neural node. It only gets much more complicated after that. Most of your learning will actually be solely mathematical and very little of it will be programming, but that is the third thing that you need to know.

Programming

You will need to understand programming to the degree of the tool that you plan to use. If you are going to a website that allows you to use a neural network that was already built beforehand, you're likely not going to need much programming. If you plan to utilize a localized version of a neural network, you are probably going to need to know how to program and access the graphics processor unit Library that's compatible on your computer. A lot of people misunderstand this requirement because in the beginning they are thinking about DirectX 11 or 12 or maybe a Vulcan architecture, but these are graphical libraries. If you plan to create a localized neural network, you will need to know a significant bit about the hardware that you plan to use. This is because you can use the central processing unit or the graphical processing unit to do the job, but how you go about using it is definitely different.

These are pretty much the different things that you need to know in order to implement and create a neural network, which is the most of what people are after when they talk about artificial intelligence. You need to know the mathematics to create the neural network, you need to know the language needed to implement it, and finally, you need to know what way you plan to implement it.

Which jobs will be replaced the soonest?

Repetitive Tasks Are the First to Go

As I have mentioned several times at this point, the first jobs that are going to go are the ones that can be repeated. Flipping hamburgers, filing, writing checks, lifting things, stocking things, ensuring things are on shelves, and pretty much anything that requires a routine. That's almost all of the low-end jobs, the ones that teenagers and the elderly tend to find themselves at. These jobs will be the first to go because you don't need to pay wages to a robot and all you need to do is maintain the robot to extract the benefits. You will still need somebody in a managerial position to handle customers, but generally, all the basic jobs can be robot.

Now, it is important to understand that there will still be one person left to just be there. This is sort of like the individual that is there at the self-checkout. The individual is not really supposed to make sure that you are checked out and get all your groceries, they are there should anything go wrong. These jobs will become the new jobs that teenagers and elderly fit instead of the ones that require the person to check out. This means that Mom and Pop shops will probably still hire the person willing to look after the register during the business hours, but a company like Walmart or Target is likely going to hire one person to manage stocking robots.

There will also be an increase in need for Maintenance Technicians and Maintenance Engineers, to ensure that the robots are properly maintained.

Jobs carried out via Rules Go Second

We've already begun seeing jobs that require rules begin to have their own version of replacements installed. For instance, as I mentioned before there is now a contract lawyer artificial intelligence that would essentially replace lawyers that focused specifically on contract work. These positions primarily follow rules and patterns, which means that even though it is significantly more difficult in routine than compared to stocking something on the shelf, it can still be automated given enough work.

Consultancy Goes Third

The last type of job that will go is consultancy and the reason why I say this is because consultancy is a routine but contextual job. Sure, you could say that in consultancy all you are doing is judging what can be added or subtracted from a workload so that the company makes more money. This is something that a machine can currently do, but the problem comes in the form of contextual understanding. You see, any machine can go and create optimization methods for a business, but the business has to create that machine to fit that business. This means that the business itself is providing the contextual understanding the business needs in order to make an effective evaluation of what is needed to optimize the business. When a person comes in to consult for a business, they need to understand the business before they begin suggesting anything. This necessity for a contextual understanding is something that can't be quantified by a machine just yet and so this is why that will be the last type of job to go. However, ultimately, it will eventually go.

Which jobs are least likely to be replaced?

Inventors

The primary job that will not be replaced, I repeat it will not be replaced is an inventor. An inventor is an individual who thinks outside of the box. They look at the market and they look at what available tools exist before they begin generating ideas for what can exist if you combine those tools. The reason why an inventor will not be replaced is because almost all companies require an inventor in order to begin a company in the first place. They are the thing that drives the industry. They absorb more data than any current processor or processor within the next decade would be able to sustain and abstract into an invention. In other words, inventors don't have any rules beyond the rules of the universe. This means that you can't automate the job because there is nothing to automate.

Frontier Science

The next type of job that is not likely going to see any form of automation is Frontier science and this is primarily due to the fact that scientists want to keep machines away from science. That isn't to say that there won't be a lot of science that these machines are capable of and it isn't to say that these machines won't be helping to march forward in the frontier science, but machines are not likely to be the entity to march forward in the frontier science. There's too much mistrust of machines, there is too much paranoia around the singularity, and if we hand over science to the machines then there will be nothing left for Humanity to do.

Will Universal Basic Income fix the problem?

Giving Everyone a Base Income

The idea of universal basic income is to give everyone a base income so that no one starves to death. This idea is not new and in fact, a lot of communist countries, as well as some socialist countries, believe in basic income for everyone in society. Due to the rapid replacement of jobs that might occur as a result of technology, many of the top leaders in technology have begun suggesting a universal basic income to offset the job loss. This would be provided on a global scale so that everyone could better their lives and it's a really good ideal but not a good idea.

Here is the concept in a nutshell because I have to describe more than what Universal basic income is as to why the leaders of Technology would believe in such an idea.. I mean bad idea. If everyone loses their job, no one has to suffer because people can still buy things if they have money. In order to ensure that they have this money, the richest people in the world donate so that everyone has a base income. This base income level would ensure that people could buy the bare necessities that they needed in order to live. This would not fix the problem of job loss, but it would significantly decrease the harmful impact that the job lost would have on the average individual because that individual would be able to buy food and similar items that would stimulate the economy.

Companies Pass Cost to Customer

The problem is that the world doesn't work like that. You cannot have a society that was previously based off of meritocracy immediately converge into a community that shares everything, it just doesn't work. I'm not saying that the idea of a universal basic income is impossible, what I am saying is that when you spend centuries building a society towards always making more and not sharing, it becomes incredibly difficult to become a community that shares everything. Universal basic income would cause companies to pass the cost of that base income to the customer by making products more expensive. The problem is that you now are automating most of the jobs that currently exist, firing employees that would have made more than the universal basic income, and now you have an influx of people who receive your money to buy your products.

Devalue the Currency

When you inflate the value of a currency, it will almost immediately depreciate in value. Let's say that we decided to put in the universal basic income and companies passed the cost on to the customer. You now have everyone at the same pay level if they don't have a job, However, money is limited. You may give them all a base pay grade, but the money is limited so unless you're going to print more money then you would have to take the money of the rich. If you take the money of the rich, the rich don't have any incentive to build new companies to become even richer if that just means you're going to take more of their money. On the opposite side, if you decided to print a lot of money you would devalue the currency. Printing money inflates how much money you have in society and this money is actually a representation of Exchange. Imagine that you had 10 tickets that you could trade in for a $300 guitar. Each of those tickets is worth $30. Now, printing money is the same as printing more tickets so let's say that you print more tickets and now you have 20 tickets for the same $300 guitar. Now, each ticket is worth about $15. Therefore, based on the example, you can pretty much see why printing more money makes money more worthless.

Everything is Now the Same but Worse

So now that you have seen the concept and you see what results from it in a very, extremely simplistic example, you can see why this is a bad idea. Sure, in the first month, maybe, the base income becomes useful but every month after that you have companies passing the cost on to the customer and then you have the worst, which is that the money that pays the cost lowers in value. This causes everything to become more expensive and the base income value is now pointless because everything you could have bought with the base income to survive is now more expensive. Therefore, everything is mostly the same but worse because now what you're getting paid is worth less than it was originally. This is why universal basic income simply doesn't work and why it has failed every society that has tried to use it.

Don't Be Scared; Use It

A.I. is just a Number

You need to understand that artificial intelligence is really just a probability matrix of what should be chosen and what should not be chosen. A lot of people think that when this reality comes into fruition that robots will kill everyone and rule the world, but the truth of the matter is that even the most complex robots that are out there today are not thinking for themselves. The rules that are in place inside of their systems are designed by programmers, which is to say that they are rules that are predictable.

All of this is just really scaled-up mathematics that has been incorporated with machines so that they are able to do the things that we need them to do. Knowing that this is something that the average person can do, it shouldn't be something that you are scared of. You should realize that almost all of the robotics makers in the world don't really have a specific intent to create robots that are designed to eliminate humans. People can't become rich off of these, people will immediately crucify any attempt to do this, and it will have a global outrage at the first person to attempt to do this. Essentially, it is one of those actions that the entirety of the world would condemn and then we would have a lot of laws that make it difficult for another person to do it.

On top of this, it is extremely difficult to create something like the Terminator. You have to realize that these robots can only work with soft bodies and we are just now exploring how we can create those soft bodies. Machines like the Terminator are significantly heavier, so something like a turret inside of a robot is not feasible because of physics. Rockets are not feasible because of physics. Essentially, if you actually look at how the Terminator is designed, you will quickly find that the Terminator that everyone is afraid of is a physical impossibility as a weapon. As an artificial intelligence that's stronger than most men, that is possible. However, having an arsenal of weaponry contained inside of the body is an impossibility if that robot is the size of a person. People are precarious designs at best and not really good at maintaining a center of gravity. In fact, this is why it is so tricky to create a human-like machine; the physics needed for the humanoid anatomy is extremely difficult to master. You might see something that comes out like MechWarrior that might have that intelligence, but you are not likely going to see a robot that looks like a human that suddenly pops out a rocket. This is reality and reality is based on numbers, if you look at the numbers it is extremely difficult to come to the fear that robots will take over in the way that Terminator took over. There's just simply so much that's wrong with the core principle that wouldn't work in our world. Don't let the fear-mongering around a robot apocalypse to decide how you want to utilize machine learning and artificial intelligence in your daily life. There are thousands of ways the world can end, the robot apocalypse is only one of them and it's not really possible.

Decisions Made Quick

It doesn't matter whether you are working in the health industry, stock market, real estate industry, or any other industry required in making decisions. Due to the fact that machines can go through decisions must faster than humans, they are able to make a much bigger change in society. They are able to quickly to decide how you can make more profit, how people can be safer, and generally decide things that benefit Humanity in a way that's faster than Humanity can decide for itself. Let's walk through an example.

Let's say that a car will fall if it falls off of one side of the building. However, once it starts rolling, you cannot stop the car. A human would have a few seconds where the human would not decide anything, and it would take them a few seconds to understand everything that needed to be decided. A computer, on the other hand, would be able to look at the scene and decide what it needed to do within nanoseconds of understanding the rules. Machines can decide things much faster than humans and can result in better improvements at a faster rate than humans can do themselves. Don't let giant companies decide the future if you can put your foot forward with barely any barriers on what you want to do.

Less Repetitive Work For All

No one really likes doing repetitive work unless it is something that is not normally seen as repetitive work but a type of hobby. For instance, there are quite a number of people that love fishing but fishing is actually a profession done by a few people otherwise we would not have Red Lobster. It's not really seen as a job until the specifics are talked about.

Artificial intelligence allows us to remove the annoying jobs that most of everyone don't want to do. No one really wants to focus on driving for 5 hours at a time, no one really wants to pick vegetables from 100 acres, no one really wants to ensure the cleanliness of the sewers themselves, and these are all things that can be handled by artificial intelligence. Don't let the fear that no one will be able to find work stop you from innovating and allowing people to take on careers they prefer more. You can decide whether there will be a lack of jobs or if the jobs that are available are jobs people love doing.

MACHINE

LEARNING

USING

FINANCE

Disclaimer

Table of Contents

What is Financial Machine Learning?

Understanding Artificial Intelligence

In order to understand what Financial Machine Learning is, we first need a brief introduction into what Artificial Intelligence is. Artificial Intelligence is, actually, a common theme in our society. From the bad guy in the movie to the interactive enemies in a game (a side note on noticing A.I. usually is the *bad guy*), Artificial Intelligence permeates our society pretty thoroughly. However, most of this is "limited" A.I.

Limited A.I. refers to a machine or program that has exceedingly limited capabilities. In fact, most of what you think as A.I. is also fake A.I. Fake A.I. refers to programming that simulates the experience of A.I. without being A.I.

For instance, when you see an enemy follow you in a game, it's not actually thinking. Instead, it's running a mathematical algorithm of "possibilities". There's a high possibility that when you are close that it

111

may melee attack. There's a low possibility that it will throw a grenade. Otherwise known as proximity estimation, most enemy A.I. in video games are simply determining an action based on distance to the player and continued accuracy of player damage. This is "fake" because it simulates a machine that has intelligence, but it really doesn't have any.

Real A.I. could be compared to the "shadow" enemy or doppelganger in some video games. In these scenarios, the game has recorded your most common actions in the past such as: most common attack command, preferred distance, delay in the attack, etc. Based on this information, this "shadow" enemy builds a fighting technique that most challenges what you are used to doing. In this way, the machine has acquired intelligence on your fighting technique. In both cases, the A.I. is limited but one has intelligence and the other simulates intelligence.

Understanding Machine Learning

Machine Learning is *sort of* like Real A.I. in that it learns based on information that it is given. However, unlike Real A.I., Machine

Learning uses something known as Backpropagation. Let's say that a "shadow" enemy is made, what happens after it is killed? Normally, a new, strong "shadow" enemy is made later on in the game depending on the length of this game. That new, strong "shadow" enemy is actually built using the same information as the first one. The only difference is that you've ultimately changed how you play and the base "stats" of the enemy are higher.

In Machine Learning, such a "shadow" enemy will attempt to see where it went wrong in the first one. This could be done by looking at the same information but in a different light. For instance, if you normally relied on a heavy attack before but switched to the fast attack rather than the projectile, it would use projectile attacks to force you to play like that. The only difference would be that now it also could look at where the standard enemies are defeated, how they are defeated, and which patterns of the attack led to more damage. If this sounds a lot like "studying your enemy", that's because it is.

Backpropagation is the linchpin to how Machine Learning works in more complex environments. It takes wrong answers it produced and finds a way to optimize for the correct answers. There are Machine Learning algorithms out there that do not utilize this technique, and these are referred to as Feedforward Neural Networks. This simply means that the data is only being fed through the system from input to output with no output becoming the new input.

In fact, if you haven't already guessed this, the Real A.I. is a Feedforward Machine Learning mechanism, but it is not a Neural Network. Therefore, while it could technically fit the category, it's not generally seen as a Machine Learning algorithm. A Neural Network is full of nodes that break the choosing factors into their most basic components. This allows the network to move fast and come to decisions in a far more accurate manner. Therefore, unless the Real A.I. is created in a parallel programmed manner, it is not Machine Learning in the eyes of the masses.

Machine Learning in Financial Situations

There are numerous instances where you might utilize Machine Learning in financial situations. For instance, you could use them to come up with a more accurate forecast based on monthly accrued financial data. You could also come up with significantly better ways to predict stocks. This could even be used to determine who is working the best out of all your workers and which workers seem to need a little more… scrutiny. This might sound Orwellian, but it's important to remember that this is still seen as mostly A.I.

However, the truth of the matter is that most of these situations are usually handled with Feedforward Neural Networks. This is because FFNN (Feedforward Neural Networks) are far easier to create and less mathematically defined. Such a benefit for this is that you might see the same information out of a BPNN (Backpropagating Neural Network) because the FFNN takes into account new numbers when it runs anew. It is actually much more beneficial in many cases to build a BPNN, but it takes more time and resources to do so. It is important here that I clarify that resources also includes data acquisition.

Let's look at the worker example to discuss why I clarify this. In an FFNN, you might want to know when the worker clocked their breaks, the clock-in and clock-out times, and maybe their quantity production level during those times. This would be a rather decently sized dataset to work with to determine high-productive and low-productive employees. However, what if you wanted to try to find the **reason** why they were low-productive employees?

In such a situation, you may want to know a little bit more on the employees; information that's usually odd to ask about. For instance, if the issue is with some of the stock personnel, you might want to know sleeping habits, BMI, and general strength. In a Neural Network, the more information you feed it, the more accurate the results *can* be. That doesn't mean that it will always be that way, some data can harm your chances. The employee might be struggling because the material is heavier than they are or they don't move fast as a result of obesity. Then as a company, you might employ free access to an in-shop gym or create initiatives to get everyone in shape so that no one is singled out, which can be a PR nightmare and just morally bankrupt.

Predictions are what ML Algorithms Do

While some may argue differently, the ultimate goal of an ML algorithm is to make predictions based on data. This means that you have a machine capable of taking data and possibly making more accurate predictions than a top economist. That means that there's also a lot that it cannot do and, ethically, some things they are forbidden to do.

For instance, you might want it to give you a 12-month forecast of potential earnings. A Machine Learning algorithm can be made for this. However, it can't tell you how to improve that number. You could create a Machine Learning algorithm that determined which tactics had better success rates, but you cannot create an algorithm that will tell you how best to sell a product. Additionally, you cannot create a Machine Learning algorithm to take over the world. I've looked it up and there are simply too many variables.

Financial Machine Learning Weaknesses

The primary issue that occurs with most of the Machine Learning algorithms is with the dataset the algorithm it is provided with. As I mentioned previously, the more useful data that you give your algorithm, the more useful it *can* become. However, a good example of when a good Machine Learning algorithm becomes "bad" is a Google example from years ago. Google is a search engine giant that makes money by providing the best search results and, thus, being able to sell ad-space that more accurately reflect those results.

One day, something odd was found in the results of a particular pair of searches. If a woman searched for a name associated with a Caucasian male, the results would often return job-related websites like LinkedIn and Monster Job. If a woman searched a name associated with African American decent, advertisements and search results of criminal background checks would accompany the original search. The problem was not how the machine language learned but the dataset the machine was given a.k.a. women who commonly sought criminal records for African American men. The Machine Learning Algorithm was making an accurate prediction, but as humans, we would consider this morally

wrong.

Developing a Trading Strategy for Stocks

There Are 4 Primary Methods of Trading

Machine Learning can actually be used to determine what method of trading you might want to take part in. Additionally, it may even help you to open up your portfolio to commit to different types of trading strategies. In this section, we're going to discuss what types of algorithms can be used alongside the different trading strategies.

Ideally, if you are going into this first hand, you might want to perform a K-cluster algorithm on the past 5 years of trading information. Specifically, you want to include all companies that have lasted for over 5 years and exclude anyone that hasn't. This is called Cluster Analysis and it will find possible associations between the companies you might not have seen. An individual skeptical of the technology will only see how the machine can show coincidences, but there is often a good reason why some stocks rise at the same time.

Let us talk about Black Box LLC (not the real company) and Lee Memorial Hospital stocks (not the real company). There's no news associating these two companies, however, what happens when LMH signs a multi-million-dollar contract with BB LLC? LMH will now be doing a lot of business with BB LLC but LMH would have had a reason to make the contract. This means that LMH plans to do well in the coming time and BB LLC will benefit from this. Once the contract becomes news, everyone will want to get a trade in before their stocks rise and this, ultimately, causes the rise. However, what happens if this happens more than once?

There's often a lot of data going around stocks and it is difficult to keep track of contract renewals. Renewals are important because they can be a predictor of rising and falling stock between those in the contract. While knowing ahead of time what will happen in the deal can be seen as insider trading, using a K-cluster algorithm to see there is a periodic correlation is not. This is why it is useful to start this before getting started as it can help associate possible relations between companies.

Same Day Trading

While I did make the terminology make more sense, it is usually referred to as Day Trading. In such a strategy, you buy stocks and sell stocks on the same day. Therefore, if you bought stocks at 9 AM on Day 1, you would then likely sell such stocks at Noon on Day 1. This makes potential, devastating losses hard to come by. However, at the same time, you are not likely to be seeing a huge payout any time soon. Due to the speed at which Day Trading takes place, it is difficult to incorporate all of the possible machine learning algorithms.

Positional Trading

Positional trading is what most people think of when they think of investing in stocks. This is usually called the "buy-and-hold" strategy as most people will wait at least a week or more after buying in order to sell their stock. This is because these people want the big payoff or control that occurs when you are a majority stockholder. This majority stockholder is actually quite popular in movies as a point of contention. However, we'll focus on the bigger payoffs like those who invested in Bitcoin for free and became millionaires a few years later.

One of the key features to Positional Trading is determining what the best categories are. For instance, a lot of people choose Technology, Law Firms, Appliances, and Healthcare right off the bat. However, those same individuals will not understand what parts of those industries are best to invest in. This is a massive problem when it comes to developing a machine learning algorithm to make a profit as it needs to understand what is a good choice and what is a bad choice.

Trend Trading

Trend trading can happen for a few reasons. For instance, one type of trend trading is when you are trying to be socially responsible. If you believe in feminism you might go with supporting only companies that have a feminist calls. Additionally, another type of trend trading is if you see a popular piece of technology that is about to come out. If you pay attention to keynote speeches or if you attend technology conferences, you might know about a certain technology you think it's going to be profitable. Therefore, you might invest in a company selling those Technologies to cash in on a particular trend.

Lastly, the other type of trend trading usually refers to when you are trying to predict the trend line for a particular company. In fact, when stock traders talk about trend trading, this is the type *they* are talking about. The trend itself is a little bit more difficult to define.

Scalping

Finally, we get to our final methodology of stock trading. Scalping is a technique that is reminiscent of the actual act of scalping. In the actual act of scalping, one takes their enemies head and slices off the skin of the skull from the scalp. In much a similar way, scalping, as it pertains to stock trading, refers to the act of buying assets at their bidding value and then finding buyers willing to buy even higher using bid-ask spreads. A bid-ask spread is simply a difference between what a stock is bidding at and the asking price for the same asset. Therefore, if you find a stock that is bidding at $40 but then you find a buyer asking for that stock at $50, you can then have a bid-ask spread of $10.

The Plan for Machine Learning

Understand How Stocks Work

"a lot of people choose Technology, Law Firms, Appliances, and Healthcare"

Ironically, those categories are usually the most stable when it comes to investing. A good handful of people say that Real Estate should be included into this, but the property has to meet specific standards in order to grow. If you buy stocks into rich communities, unless that community is expanding, that community is likely to depreciate over time. The reason why those categories are the best is usually due to two factors; P/E and P/BV ratio.

In places of technology, you have a wide range of P/E. Let's get something out of the way real quick because shares and stocks can be quite confusing on their own.

A Share is how much the company wants to divide itself, which is announced by the company itself. Therefore, when a company goes "public" what they are doing is opening the Share pool to the public. At this time, they will release a specific number of shares. Essentially, they are selling components of a company to the public as a stock in the company. This stock is called a Share, as in you are getting a share of the company. A stock, on the other hand, is a collection of those shares. It's rather hard to keep track of 1 billion shares but if each Stock is made up of 5 shares, that's only 200mil stocks to keep track of. However! This is a story of all shares are stocks but not all stocks are shares. For instance, some stocks could be allocated to certificates and other parts that are share-like but are not directly shares. So, let's begin this by understanding how to calculate for shares since they hold the **real** value for stocks.

Let's first discuss BV or Book Value and to get this we need to break a company down into its most basic components and go from there. Let's go ahead and define asset and liability. An asset is anything that holds monetary value, whether it makes you money or you can sell it for money. A liability is anything that holds a deficit of monetary value, whether it costs you money to keep it or if it is a debt.

Let's say we are a software company and we're about to open up to the stock market. We know that we make $100 for every software we make, so let's use that as our basis. That $100, should we only sell 1 product, is called the **Total Revenue**. However, we only got that software because we had it made. If we have a software team and website, along with other items, that we pay for then this would be referred to as the **Cost of Revenue**. Let's say that for every product we sell, we have a Cost of Revenue of $70.

$$Net Income = Total Revenue - Cost of Revenue$$

This number is called Net Income *and* is called **Earnings** when divided into shares. If there are 10,000 shares and you own 1 share, you make 1/10,000 of the Earnings as a result. This is the reason why finding out what a company is worth before you buy a stock is important. This is why you need to know the **Income Statement**, **Balance Sheet**, and **Cash Flow Statement**. You already know about the Income Statement as this is where you find the Net Income and Earnings. The Balance Sheet is where you'll find the Equity and Book

Value or BV of a company. This will be where your **Margin of Safety** is.

$$Equity = Total Assets - Liabilities$$

That equation is what calculates the Equity of a company, but the **Book Value** is a little bit different. Additionally, if you take the Market Price, the price that the market is selling all combined shares for, and subtract the Equity from it, you then have the risk. Therefore, if the Market Price for this software company's shares are a total of $1,000,000 and the equity is $20,000, then you have a risk of $980,000. This would be referred to as an extremely low Margin of Safety. While this bigger number is known as the Equity, when it is divided into shares it is renamed Book Value.

P/E refers to how much the company's market price is versus how much it earns, thus aptly named Price to Earnings Ratio. There are a few steps needed to get this number. I'm going to not only explain how to come to those numbers, but why you need to know them even though most stock "news" websites provide this freely.

To get to the P/E ratio, you must first find the EPS and that simply means how much each share will earn you should you buy it, named Earnings Per Share. EPS is calculated by taking the Earnings and dividing it by the Outstanding Shares number, which is really just how many shares are in circulation. To find Earnings, stock traders will take the Total Revenue of a company and subtract the Liabilities of the company. Therefore, the true calculation for this is:

$$EPS = (TotalRevenue - Liabilities)/OutstandingShares$$

The P in the P/E ratio is the current price for each share as it stands currently. The EPS is how much each share should cost as its' base value. Therefore, if a company says it's going to sell 2,000 shares at a total value of $10,000 then the value of each share is $5. This is the P side of the equation.

If we take the $100 for our software company earlier and turn it up to $10,000 and we do the same with the $70, which becomes $7,000 we can now do EPS calculations. This means that our Cost of Revenue, which is also our Liabilities in this situation, turns each share into $3000/2000. This makes our EPS 1.5. Now, to get P/E we do:

$$P/E = MarketPrice/EPS$$

Therefore, if we say that the Market Price is $5 and our EPS is 1.5, then our P/E is 3 ⅓. This is important to understand because this means we need 3 ⅓ shares to get us an annual amount of $1.

So, now that we have worked through this, you understand the general gist of most of the stock market. Good investors will look for a low P/E value and a medium to high Margin of Safety. While there's definitely a bit more to stock trading, this book is about Financial Machine Language and how it relates to the stock market.

Why You Need to Know How It is Calculated

Information is Not Real-Time

The most important part of this is that you needed to know the previous information because even if you rely on the information gathered at stock market APIs, they may not have the most recent, updated information. For instance, a common trick to see if the information is updated daily or updated in real-time is to check the P/E value for yourself. Sometimes, the number for this could be one or two points lower or higher than what you personally calculate, which means what you calculated is more accurate than the information you're being provided with.

You Need to Teach the Machine This

You need to actually teach this to the machine, so all the equations you would normally do in a financial situation needs to be done by the machine itself. After all, what is the point of building a machine learning neural network if the neural network cannot function in its original purpose? You can use machine learning to identify Trends and similar, simple tasks. However, if you're going to spend the time to train a neural network, it might be worth your effort to build a machine that could suggest trades rather than look at common market trends and just provide general data you already have access to.

You Need to Know Where to Point the Machine

No matter what you are doing with a machine learning algorithm, you are going to need a way to point it in a certain direction. Whether you plan to use the machine to identify trends, to identify companies worth investing in, to see if there are hidden relations between companies, or anything else, you need to at least know the mathematics used in the stock market so that you can effectively point the machine towards the direction you wanted to go.

Understanding How a Trader will Trade

Now we get to the **point** of *this* chapter, but we first needed to set up a base to understand this. When you develop a machine learning algorithm for financial purposes you need to know what you will be using to teach that machine learning algorithm. If the machine learning algorithm doesn't know how to make the proper Financial choices, it doesn't make sense to make a machine learning algorithm or try and use one to replace your own decisions. Therefore, now that you know how the different stock strategies work and the math that goes along with the stock market in general, we will now go over the different parts you need to build in a machine learning algorithm.

Point-of-Direction

While most might think that the Point of Direction is "most profit", but this is not how a stock trader thinks. In fact, "most profit" is a benefit of what they actually think of. Most stock traders will focus themselves on undervalued assets, currently expensive trends, and what's currently selling for fair value.

There's really a benefit to each of these, depending on how you handle them. For instance, the easiest one to think about is finding undervalued assets but it is the most difficult to resolve. The previously mentioned method in Positional Trading is actually referred to as the Warren Buffett way of trading. You calculate what's currently undervalued if they have a low P/E and BV. This tells you that you will get a lot of money for anything you purchase, but there's more to this story. For instance, what if you were trying to find an undervalued new piece of science?

It was generally thought of as a hobby practice in the 1960s. Only Rich individuals could really invest money in this and you had to order it through very selective magazines. In addition to this, those magazines usually funded the project themselves. This was a fairly complex item to put together and normally it wouldn't make anybody any money because the parts were so expensive. Now, would you invest in this hobby? If you said no, you would have been part of the crowd that didn't profit off of the future of computing. This was a fairly simple example of when undervalued technology enters the marketplace and people don't see the potential future. More importantly, how exactly do you put a value on such an item beyond visionary? The truth of the matter is you simply can't.

The easiest to define is actually allowing a machine learning algorithm to spot currently expensive trends. For instance, how would you identify a trend? You would likely look at the previous data and see that there is a threshold for an upper bounce and a threshold for a lower bounce. This can be easily defined inside of a machine learning algorithm because thresholds are usually what it works in the first place.

You see a particular stock begin to rise and then fall and then repeat this pattern for a few steps, which ultimately leads to a great rise and then a great fall. This is usually referred to as a trend where the company gains a little bit of value and loses a little bit of value but every once in a while, it gains a lot of value and then loses a lot of value. This practice is known as Trend trading and it is fairly common for people who want to make a lot of money quickly but not spend their time watching the stocks on a daily basis. Sure, Trend trading can happen on a daily basis but when you're working with certain companies, these trendlines can actually take weeks to develop. It's not really something that you can predict beyond past patterns of how fast that happened in the past and what caused those events.

Finally, you could simply set the machine learning language to identify sections where the amount is of a fair value and that is to say that it is about equal to what it should be. This is actually pretty hard to do because once a company gets pretty popular, this fair value tends to be thrown out the window. A perfect example of this is the Bitcoin as it is worth far more than what most people would generally say it's worth.

In fact, it's usually referred to as a new type of Fiat money because of its initial value and then it's true value being so different.

The problem here is that I can't tell you exactly what point of Direction you should use because it's too difficult to determine the best move. There are better moves that you can make. For instance, you definitely want to try and find the cheapest type of asset that you can get and you definitely want to make sure that you trade in for stocks at a decent value. However, that limit and benefit is really defined by you.

Entry Trading Price

The entry trade is really just the name you use to describe the moment you would find a stock acceptable to buy. In such an instance, you might want to buy stocks at around 20% of the value that they are currently being sold at. This would be a threshold that you would set inside of the computer to determine that you want to buy it at a certain price. The problem here is that you run into an infinite possibility issue. Let's say that you set it to only buy stocks if they are 20% lower in value than when the stocks first opened. Sure, let's say that your machine will now do this and that you, in fact, receive a windfall of stocks because the stock value has gone down by that much, but what happens if it goes down another 20% because the stock market value is crashing for that particular business? That's a huge risk to take and yet it is a very real possibility. Even worse, this could have been any percentage that you said ahead of time because even if you said it to 5%, it could very well go down another 5%.

Additionally, you run into the other side of the problem, what happens if it never goes below that amount? You would then need to adjust it to the amount you felt appropriate, which would likely be based on the lowest common price that it drops to. However, you run into the issue happening again. It is rare, but it does happen, and these are the things you need to think about when going with machine learning and determining your prices for your stocks. After all, you don't want to buy a stock that is plummeting into the ground and you also don't want to have a machine that will never buy any stock because it never reaches the threshold you have set. However, the Entry Trading Price represents the *identification* of a buy-worthy item.

137

Exit Trading Price

You're going to need to define a trading price in which you exit the market, which is to say that you need to find a price that you are comfortable selling at. Once again, you come into the problem of trying to determine a value that most people determine at the time of looking at the stocks. However, we actually do this all the time in stock trading. This is often referred to as stop order and limit order. For instance, in a stop order, you may only want to buy or sell an item at a specific amount. Until that amount is reached, no money leaves your pocket. In a limit order, we are dealing with the same thresholds a machine language will have to deal with.

A limit order is a type of order that allows you to set a minimum or a maximum for which you will buy and sell at. These are usually speculative numbers that you think the price of the item will reach in a given time period. For instance, let's say that you decide to buy Facebook stock at around $70 per stock. You think, according to pass data, that there is a likelihood that your stock will reach a total of $75. However, you will see that it is possible for your stock to become worth

$65. Based off of your trading strategy, if you invested in Trend trading, you might think that a big Surplus will likely lead to a huge debt. You can then set a limit order to buy the stock at specific quantities, which we will talk about in a little bit after the price of the stock has gone below $65. Likewise, if you do see your stock become worth $75, you can set a limit order to sell that stock once it goes past $75. As you can see, you can set the thresholds of the machine language algorithm to utilize limit orders much the same way that the standard stock trader might.

The Exit Trading Price represents the *identification* of an item worth selling. It's fairly important to recognize that when dealing with machine languages, you have to identify the boundaries before you can use them. With the entry trading price, you are finding products that are worth buying. With the exit trading price, you are evaluating items you have already bought and determining whether it is time to sell them or not.

The Tolerance Range

The reason why I pointed out that the entry trading price is identifying items that are worth buying and the exit trading price are identifying items you already have that are worth selling, is to modulate the machine language algorithm. In both instances, you may follow similar steps, but the data set is significantly smaller for the exit trading price and the mathematics are in reverse to that of the entry trading price. The purpose of these two variables are to identify the upper bound threshold and the lower bound threshold, which requires two separate lanes of logic and data sets.

The Quantity of Sale

Finally, even though it seems obvious, I believe I need to ensure that you understand that one of the key components of any machine learning algorithm dealing with stocks is how to determine how much to sell. While the new stock trader might think that it is a good idea to sell all of their stocks once those stocks become profitable enough, that isn't always the case. For instance, let's say that you have 1,000 stocks inside of Intel. If the stocks rose by $5 then you would have a profit of

$5000 if you sell those. However, if you plan on seeing a possible $20 P/S return, you may only want to sell a few of them while they are profitable and wait to see if the price increases further. After all, sometimes there is an increase in the trend that will eventually lead to a greater downfall, but you can still make quite some money if you look for a climbing trend. Therefore, you may just want to sell 500 of your stocks so that you at least get $2,500 as a profit and then wait to see if it falls back down, which you can then use the profit to buy the stocks at a lower price while waiting for the stocks to climb back up. This allows you to make more money while also keeping reserve stock for when the price is even higher. This means you could have more than one neural network handling portions of stock with different limit orders on them.

Machine Learning to Determine Value of Current Stocks

Training Your Machine Learning Algorithm Quarterly not Annually

Now, you might think this is a bit odd, but when you begin to think of long-term investment, you should really be looking at quarterly reports. However, a normal stock trader will not look at quarterly reports unless they absolutely have to. This is because the quarterly reports can usually be summed up in the financial statistics provided by their chosen stocks' website, which are much easier and quicker to look at. Long-term investment means that you need to find a safe pattern with a specific company to see if it is within acceptable ranges of doing business with.

For instance, as I mentioned before, a company with a great PE value is usually a company that's worth looking into. You then look into the company and find that they also have a great PBV value. This means that the company might be worth investigating, but the problem is that the PE value changes over time. While you may be looking at the current PE value, the past PE values also have, ironically, value. You see, the PE value and other variables can be looked upon over the course of time to determine the overall success of that company. Ideally, you want companies that have the same value or similar value overall.

This is because a company that is worth investing in for the long-term has to have a stable monetary environment. Once you look at the PE value, you might also want to look at the equity that the company has held over those quarterly amounts. While the PE value can tell you how much money you will make by investing with such a company, the equity will help determine the viability of the company. Additionally, a company that may have a decent PE value might also not be able to handle their debt very well or invest back into the company so that the company makes even more money. These are all problems dealing with long-term investment and finding the best companies means that the machine learning algorithm needs to also know which companies in the dataset died and which companies in the dataset lived.

The reason why knowing whether the company is still around or not determines a pattern of death. While it may be somewhat not surprising, companies tend to die in very much the same way. Normally, what will happen is that the company will suddenly have a negative value in their revenue. Now, they could also sell all of their assets before the negative value appears or afterward if they weren't expecting it. This key moment represents when a company is panicking because then the natural pattern is to consistently take on more debt as the company just dwindles into death. On a rare occasion, they do survive, but more often than not most of them will die in very much the same way. Understanding that debt and total revenue play a key role in determining when a company is about to die is key to teaching Machine Learning algorithms to stay away from companies that exhibit the same patterns on a quarterly basis. So, your Dataset may look something like this

Date	P/E Ratio	Total Revenue	Debt	Asset Value	Liability Value	-1 = Dead 1 = Alive

The Common Sense Method: Trend Trading

The most common method of trading is to use peaks and troughs, which is really just to say that one would follow the Trend. Whether you are looking at the Long Term or Short-Term Investing, you are still looking at Trends. In fact, early on I mentioned that there are 4 primary methods of trading, but in reality, all of them still fall underneath Trend Trading.

Let us take a look at the following chart for the Dow Jones in June of 2018 thanks to CNBC:

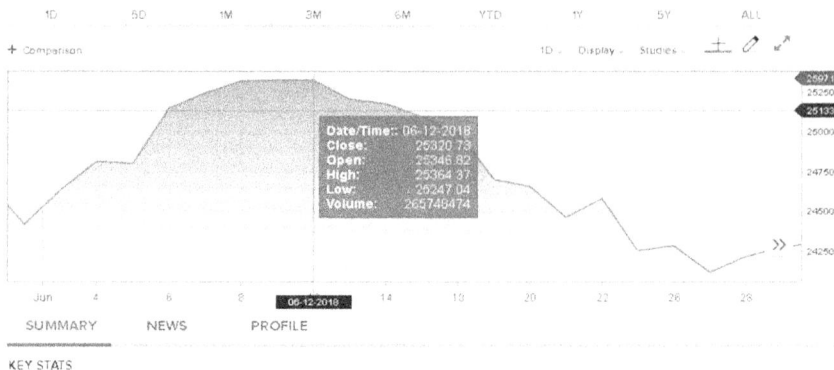

As you can see, this could be one of the better or worse months depending on which side of the hill you sold your stock on. We see two specific type of trends in this chart, I will note them as Green and Red lines in the next chart:

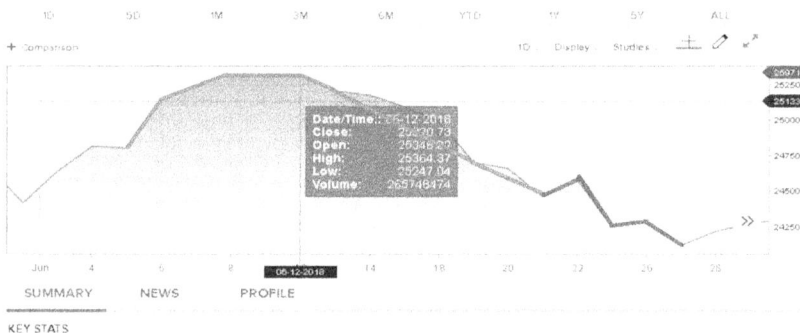

Before, we discussed how to find businesses that were worth investing in during the *Understand How Stocks Work* section. Once you

146

have found a business that's worth investing in, there are 2 different trends all good businesses will experience at different intervals. The Green Line is known as the Big Peak, which is to say that the line will Peak before massively falling. This is the part of the line that people want to sell on. The falling part is known as the Trough Line, which is the area where people wait to buy again. The Red Line is the Little Peak, where most Day Traders spend their time. It's the same as Big Peak, but profits come at a much faster pace. If we look at a 6-month version of this, we can easily see quite a few of these:

This, for us, is fairly easy to understand and becomes a repeatable pattern we can look for. Provided we are looking at successful businesses, the models are pretty straightforward and don't require a lot of math unless you want to become super precise.

The Most Common Machine Learning Method: Support Vector Regression Algorithm

It is not as simple for Machines to use the same method of analysis. This is because we have already constructed the algorithms needed to recognize this type of patterns. In fact, the human brain is genetically predisposed to recognize patterns. However, creating a Neural Network is like teaching a toddler how to recognize the same pattern, only machines catch on quicker and require mathematical explanations. Therefore, we have to separate these patterns into two different categories.

The first category is to notice the extremes of the lines, but which ones. Since our brains already recognize the patterns, you might be stuck on thinking of the line, but you need to stop that if you are. The Machine does not see a line, instead, it sees a sequence of positive numbers and we need to retrain the Machine to see something else. It needs to see that positive numbers closer to zero are a **Loss** and positive numbers further away from zero are a **Gain**.

However, you've probably already noticed at there's no good "line" that fits the model. If we look at the previous data, when you define a line for one trend, it will no longer be applicable for another trend. Therefore, we need a way for us to continuously define where the Losses and Gains are on a Line, which is where Support Vectors come in.

The Support Vector algorithm sets a margin for Gain and a margin for Loss, which are placed along a Trending line. So, let's use the previous definitions for Gain and Loss. Definitions for these lines are a little bit different though. Instead of going from zero, we go from what's known as a Hyperplane. This is a separate line that places itself between the D+ and D- margins, but these margins are defined by the closest or furthest vectors based on your needs. One category is D+ and the other is D-, based upon what you want. These are just classifications and D is just a placeholder, here is the following graph that

demonstrates how Support Algorithms would be displayed:

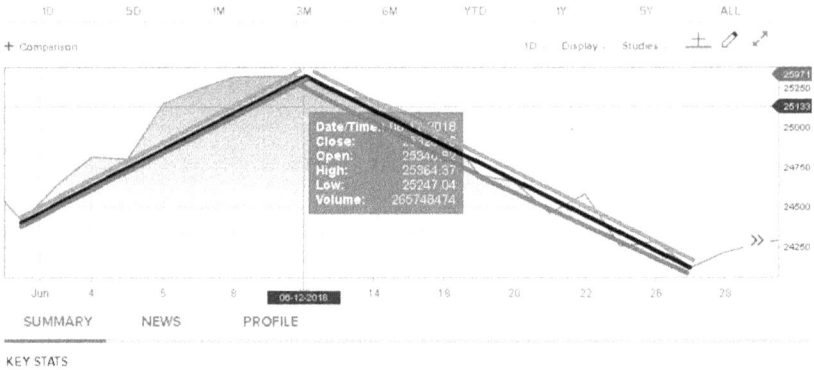

As you can see, the Green Line or D- in our case is the Gain

line. The Red Line, on the other hand, represents our Loss line and is

the D+ of our equation. Now that you understand the general gist of

how it works visually, we'll give a brief explanation of a massively

complex algorithm you're going to need to understand what most

machine learning stock prediction algorithms will use:

$$y = \sum_{i=1}^{n}(a_i - a_i^*) \cdot \langle \phi(x_i), \phi(x) \rangle + b$$

Don't freak out. It's okay. We're going to walk through this together. You need to understand that this is actually a combination of 2 algorithms. Additionally, this is just an evolved form of

$$y = mx + b$$

Almost all machine learning algorithms start out with the Slope-Intercept and this is because the Slope-Intercept is the first *true* form of Binary Separation. In other words, Slope-Intercept is the same type of formula used in the Perceptron. So, let's go ahead and break this information down into the different components:

$$m = \left(a_i - a_i^*\right)$$

$$x = \langle \phi(x_i), \phi(x) \rangle$$

And I think we can understand that b = b in this situation. So now we're going to break this down even further, but we'll start with x because it is less abstract than m. Let's start with the most confusing part, the angled brackets holding this equation together. You will find these angled breaks when dealing with **Linear Span** and **Linear Algebra**. To understand Linear Span, you must understand **Linear Combination**. A Linear Combination is a combination of Linear Vectors, which is to say if we have a vector of 1 and a vector of 2, this produces a vector of 3 otherwise written as:

$$3 = \langle 1, 2 \rangle$$

Now, the way we would normally write this is

$$\vec{3} = \vec{1} + \vec{2}$$

I would say this makes for an efficient explanation of what's happening with the variables inside of that equation. So, let's rewrite this so it looks a little bit more understandable:

$$x = \phi(\vec{x_i}) + \phi(\vec{x})$$

That little arrow on top represents this as a vector, but now... which vectors and don't vectors come in pairs? Let's address the "pairs" theory because you are technically right but let's think of this in slope form. If you recall, in slope-intercept form, m is a slope, x is x and b is y-intercept. Therefore, *a* actually refers to slope here, *x* is not a pair but a single number, and *b* is still y-intercept. Additionally, one can write (3,1) as x and y, but it can also be written as $(0,0) + (3,0) + (0,1)$ where (0,0) is the origin. Therefore, Vector X is still 3 here, it's just been separated from y and origin is assumed because of the pairs, the zero is the origin and the second is the direction thus (3,0) or 3.

However, in order to understand "which vectors", we have to look at the part of the original equation we have not discussed:

$$\sum_{i=1}^{n}$$

Now, you might be confused by this, but this is really just the mathematical representation of a **for loop** in programming. The $\sum$ part is the Sum of all ns starting at i. Therefore, i represents the current iteration number in the sequence. Just to let you know, this is the **regressive part** as this allows it to iterate through all the vector points in the past by starting at the earliest dated variable. So now we can understand that $\phi\left(\vec{x_i}\right)$ is actually Phi of the current vector for x.

We have explained x so now we need to handle m, which is a little bit more difficult to understand. The part that throws most off is that little asterisk in the corner:

$$\left(a_i^*\right)$$

That asterisk is known as a complex conjugation, which should really be called a reverse or an opposite function because $\left(a_i\right)$ is the opposite. In programming… we would just multiply by -1, am I right? I am **wrong**. It's important to understand how conjugates work. Let's say I have the equation:

$$y = 3x + 2$$

If we multiply everything by -1, I come up with

$$-y = -3(-x) - 2$$

This is **NOT** how conjugation works. In conjugation, I would end up with:

$$\bar{y} = 3x - 2$$

This is an entirely different equation, right? That's because we're locating a number that brings the imaginary number to its' opposite self. If I said x is 1, here's how this would go.

$$\left(y = 3(1) + 2 = 5\right) - \left(\bar{y} = 3(1) - 2 = 1\right) \not\equiv 0$$

Here is the official conjugate formula:

$$x = a + b_i \ \bar{x} = a - b_i$$

In this equation, we have 1 imaginary number and that is b_i where b is the Real number and i is the imaginary number. Therefore, we can only apply the conjugate if the equation looks like:

$$Actual(y = 3x + 2)$$

$$Conjugate(\bar{y} = 3x - 2)$$

So, in the original equation of $\left(a_i - a_i^*\right)$ what we are actually seeing is a Real Number of iteration i minus the Conjugate of the Real Number of iteration i. The last part of this equation is to actually substitute something known as a **Kernel Function**. Now, the most common of these are Polynomial and Gaussian Radial Basis.

Polynomial

$$k(x_i, x_j) = (x_i, x_j)^d$$

Gaussian Radial Basis

$$k(x_i, x_j) = exp\left(-\frac{\|x_i - x_j\|^2}{2\sigma^2}\right)$$

These Kernel Functions allow for Non-linear SVR, which is what we see in the stock market. When using them for this purpose, we use a Kernel Function to replace this

$$\langle \phi(x_i), \phi(x) \rangle$$

If you would like to study SVM more in-depth, here is a great series of web tutorials: https://www.svm-tutorial.com/2014/11/svm-understanding-math-part-1/.

Determine the Optimal Time to Buy Stocks

You Should Not Fully Automate

Now that we have gotten off the beaten path of practically pure mathematical reasoning, I need to tell you upfront that you should not fully automate your trading decisions. There are some that are willing to convince you that fully automated trading is a fantastic idea, but the honest truth is that it's a really bad idea. It's a bad idea because we, as humans, are definitely flawed. What I mean by this is if you didn't train it right, if you missed a line of code, or if something goes wrong in the processing speed, you could make trades you would never make in your entire life.

The problem with developing a machine learning algorithm that's fully automated is that you have to be able to trust that it's going to make the right decision most of the time. The problem with humans is that we are often way too impatient to wait until it can get that good and so we try to use it almost as soon as we build it. The problem with

158

this is you can wind up broke and out of the game within moments of you turning on the machine learning algorithm if you did something wrong. Instead, you should really be using it like a personal advisor at times. Specifically, you should know more than it for a little bit down the road after you have created it.

The moment that you can feel comfortable handing over your trading decisions to the machine should be when it gets the calculation right more than you do. Therefore, say you were to get signals from the machine learning language algorithm that told you that you needed to buy stock and you refuse to do so because it didn't make sense to you, but it turns out that the machine was right. If this happens more than 75% of the day, I would highly recommend just leaving the trading decisions up to the machine because at that point it's more successful than you. However, it's going to take quite some time for it to get up to that type of accuracy and a lot of practice. I would not trust the machine to make any trading decisions within at least the first six months of the machine's creation.

Using Support Vector Regression Analysis to Find Good Times

While Support Vector regression analysis is definitely a fantastic technique, you have to remember that it is it's only designed to predict the next step in the Trends. Therefore, it can actually tell you what times are going to be good for selling and buying your stock. However, it can only ever predict the next step in the process so if it's wrong then it has to recalculate everything until it becomes right. Considering the sheer randomness of the market, it's going to be recalculating quite a bit so while you can use it to find good times to sell your stock, it's not going to be 100% reliable.

We have not gotten to the point where machine learning is more efficient at humans at doing particular tasks 100% of the time. There are certain machine learning applications that can claim that they do it better than humans most of the time, but none of them have significantly proven to be flawless compared to humans. This means that you can utilize it to try and find times to sell the stock in real time, but there are a few other methods out there that can help you determine great times to sell stock.

Specifically, I'm talking about partitioning cluster methods. Partitioning cluster methods are clustering techniques that allow you to separate information into clusters. Therefore, if you are able to find profit spots more than loss spots, you can actually see a pattern of when it's good to sell a certain stock with a company.

For instance, let's say that we divided our information into 30-minute increments for a specific business over the period of a day. During that time, we record moments where the company is at a profit and moments when companies are at a loss. We can actually do this in a line graph, but clustering techniques are specifically meant to grab data points that form into a cluster. Therefore, while you could see a trend line in said 30 minutes, you are more likely to see dots of profit and dots of loss. Therefore, as a natural result, you can actually find the parts of the day where there are more profit dots more often than there are loss dots.

Using Cluster Analysis to Find Good Businesses

Half the battle in determining what companies to sell with and at what times you can sell at is often the business itself. You don't really want a business that is dealing with a massive pile of debt that it can't seemingly pay off. You also don't want a company that if it died yesterday, you would likely only see a small portion of money go back into your pocket. While support vector regression analysis is really good at finding the trend times that you want to sell your stock at, you primarily want to use partitioning and cluster algorithms to determine which companies you should go with.

While we will get into specific clustering methods later on when we discuss penny trading and day trading, clustering algorithms allow you to group together common companies. Therefore, let's say that you have companies that do extremely well, companies that do generally well, companies that do okay, companies that are losing a little bit of money, companies that are losing a lot of money, and companies that are about to die. Now, there are also over a million of these companies and you can't possibly check them all within a reasonable amount of time. Normally, you would generally just look for the best ones that you can cherry-pick out of the data set. This is a problem because you could have missed all of the ones that could make you money, it just gives you the ones that are most popular because, as humans, we are very biased. By using cluster algorithms, we can use the variables that we would normally use to identify companies in those categories and then cluster them on each variable so that we can compare those groups rather than individual companies.

This allows us to group companies together with other companies with the same traits and values without spending a lot of time looking at the data ourselves. To do this, you don't actually need any machine learning algorithm, you need cluster algorithms that take data and make relations to it. This is actually the first step utilized in machine learning to create labels that the machine learning algorithm can then use.

Using this for High Volatility Stocks

This method is actually still pretty useful when considering it for high volatility stocks. High volatility stocks are usually the bane of the existence for the average stock trader, primarily because of how unpredictable they are. The problem is that it has been proven in the past that high volatility stocks can be predictable if you look in the right places. High volatility stocks are almost unanimous with penny stocks and it's because of how penny stocks work. Therefore, primarily you want to look for the same aspects that you would look for in a penny stock business as you would with normal high volatility stocks because you'll find that the similarities are very apparent. That isn't to say that there's not a difference between high volatility stocks and penny stocks, but you look for the same parameters for both because that's usually how you assess whether those stocks are going to make you money or not.

Machine Language Algorithm to Predict Whether to Sell a Stock

Trade Signals are Safe

The safest way for a machine language algorithm to help determine whether you should sell a stock or not is to set it up to give you trade signals. This works in very similar ways to the stop orders and the limit orders that people use to take advantage of automated trading but with the ability to make the decision yourself. For instance, you can get a trade signal that tells you that a current stock price is worth selling at that time. If the machine was allowed to make the trade for you, you would sell your stock at the moment the signal would normally go off. The reason why I think this is crucial to understand this is that you could make more money if you waited longer after the trade signal.

A normal stock trader will have one of three options when dealing with an automated trading system. The first option is that the trader doesn't use the network at all, this is often referred to as the imaginary choice. The second option is that the trader uses the machine language algorithm to give the stocks trader advice. The last option is for the stock trader to fully trust the machine to make trade decisions on behalf of the stock trader.

Having a Machine Trade For You Pros

There are a few benefits that come with allowing the machine to make decisions for you. The first benefit that I would say that machines have over humans is the ability to make that Split-Second sale. There is a period of time where the user may take longer to make a stocks decision, and this can actually lead to a loss of money if the decision isn't made fast enough. Additionally, this lends to the fact that machines are faster at analyzing information than humans and that's simply because they are built to only do that whereas humans have a whole bunch of mechanisms for survival that get in the way.

This actually leads this to the secondary benefit and that is really that machines don't have emotions. For instance, a person will probably never invest in a company that they hate no matter how profitable that they may be. That isn't to say that the person doesn't want to make money, it's just their emotions are getting in the way of making the money. Machines don't have that problem and all they care about are executing the tasks that are given to it via programming. Therefore, machines are more open to the types of trades that they can make whereas humans purposely limit themselves based on their emotional background and their history.

The last benefit that I would say that machines have over humans is that they can analyze bulk information far quicker than humans. This is different than the split-second sale that I was talking about before. A lot of information goes in beforehand to determine whether a company is worth investing in or not and humans have to look up the numbers and do the research themselves. A machine that has access to an API of information will actually be able to analyze those numbers based on the parameters that the human set beforehand. While it may take a human nearly a week to study and determine the worthiness of companies that match his profile, it may take a machine an hour or two at most. This makes them highly efficient at finding good companies quickly and then putting into action good policies to make their owner's money.

Having a Machine Trade For You Cons

There's one huge disadvantage that comes with having a machine trade for you and a lot of other little disadvantages. The one huge disadvantage is that once it starts to mess up, it cannot stop itself. A machine is designed to continuously run and if it makes a wrong choice and the mathematics are devolving, it continues to devolve into whatever mess is you wind up catching it at. This means that you could be a millionaire and if you let a machine control your money and that machine begins to mess up, you may not catch it until you're broke. This is a huge problem that belies all types of Technology but especially machine learning technology. This is why you have to spend months testing the software before you can trust it. You have to constantly make sure that your choices would line up with the machine's choices and vice versa.

This actually brings me into my second point and that is that it takes a lot of time to make sure that a machine learning algorithm is up to par with what you would do. You need to take time out of your day to invest in the machine learning algorithm so that you can test it, correct it, and perfect it. It takes months and even years to make a machine learning algorithm that won't eat itself when it starts to mess up and will provide you with accurate information. This is why most people leave it up to big corporations that make money off of this software because they allow those corporations to invest their time and the person paying them is saving time by allowing the company to do it for them.

Now, there are a whole bunch of other problems that come with developing a machine learning algorithm. I won't list them here because I could actually write an entirely different book on the problems that can happen with machine learning, but there is a litany of issues that a developer has to deal with when handling machine learning algorithms. Yes, most people will brag about how machine learning is awesome and how machine learning can basically solve every problem on the planet, but the machine learning itself is a very tiresome and difficult science to put into practice. It takes a lot of patience but, usually, the payoff is generally worth the amount of work required to put in.

It's Really All about Predicting the Trend

The last thing that I want to say here is that this is really all about predicting the Trend. Every single thing in this book is really just about predicting where the trend line is going to be and how you can use machine learning to predict that trend line. That Trend line is how we make money, which machines are really good at predicting because we use mathematics to predict it ourselves.

Determine Value of a Penny Stock

Long History Data is Not Good

Ironically, a penny stock is one of the very few stock options that you don't want a lengthy history about. The reason being is that penny stock is usually penny stock because either the company just started or the company is barely on the stock market radar because of reasons. Often, I find that the reason why many companies are on the penny stock list is simply that they were worth a lot of money and then they got into a lot of trouble that eventually plummeted their overall worth. Most times, I've seen companies go from $60 or $100 all the way down to $3 within the span of maybe a year.

This means that if you were to feed in the information to the machine learning algorithm from a long history, the machine would learn of the huge losing gap that occurs with penny stocks. In fact, if you based your profit margins off of past iterations, you might actually see that you may never buy penny stock as a result. Now, most of the penny stock makes you money through bulk purchases. That is to say, if the stock is worth somewhere around $0.27, if you buy a hundred of that stock, you now have around $270 in that stock. If that stock goes up by a single penny, you then go up by $10 as a result. This is where the money is for penny stocks; bulk orders of the stock because any fluctuation towards the positive usually means you're getting massive benefits.

The problem is really determining whether a company is dwindling and dying or if the company might make a comeback. It's kind of sad to say, but the penny stock is really the market where the average stock trader is kind of rooting for the underdog at that point. These are companies that have been at the height of their game and something happened to where they're almost next to worthless on the stock market. The stock traders in this stock market are using and taking advantage of the cheap prices to make multiplicative dollars off of pennies, which actually helps keep those companies alive. Remember, when the stock goes up, it usually means that the actual value of the company goes up as well. Therefore, successful trading in stocks with the company that is in penny stocks might actually bring them back to their height of glory that they once knew.

Beside that sentimental view, you are defining the successes based on pennies and not dollars, so your thresholds need to take this into account. If your stock originally started at $70 5 years ago and is somewhere around $0.50 nowadays, you are not looking to make the $70 that that company once had. You don't want to overload your machine learning algorithm with useless information. This would mean that you are able to train your machine on data sets quicker because it is only data that is the most useful. Likewise, you really shouldn't need more than a year or two to gauge the success of certain penny stocks.

History of Debt is Good

On the flip side of things, just as you don't necessarily want a long history on the company and its profits, you do want a history of debt. You see, one of the key factors inside of penny stocks is whether that the company is dying. The one thing that you look out for is a company that is surviving and a company that is continuing to die. You

173

see, a company that is surviving will eventually rise back up and that's the ultimate wish for a penny stock trader because if you have a hundred stocks that you bought at $0.27 and the company rises to a stock value of $30, you spent $270 to make three grand. That's a huge amount of money but it's usually way too risky to do that. Instead, what you do is you look for the trends of the day.

Just like with most trading, you want to look for Trends in the data itself and you want to make sure that you are only looking at trends for companies that are surviving. This means that you don't necessarily want to look at all the companies you can inside of the penny stock market. Instead, what you want to do is you want to filter out all of the companies that have an unmanageable debt. The way that you do this is you collect information on equity for all of the companies. You then set up a margin that you find acceptable for equity levels. You can also throw in total revenue to see if they are surviving by subtracting the total revenue and the equity, but essentially you want to make sure that you capture the debt and you want to pay attention for debt that is decreasing. Decreasing debt means that the company is starting to come back. You won't necessarily see a revenue spike because the company is struggling and is probably making the same if not a little bit less than what they're used to.

This debt history can actually be used to calculate the company's margins of risk. Therefore, if a company has a massive amount of debt and it doesn't seem to be lowering, the machine can use clustering to put that debt area into a specific graph that's away from the others. You could also have it separate the margin of risk by percentages so that if it fits within a certain percentage then the clustering can bring that into groups so that you can easily identify which ones have high debt that's being managed and which ones have low debt that are doing okay and then you have ones in the middle that aren't able to seem like that they're getting rid of that but the debts not increasing. What this does is that it creates categories of risk that you can easily identify as companies worth investing in. The reason why you might want to use clustering for this is because of the sheer amount of data that you have to go through. You may have to only look at a single number, but you also have to look at a single number 8 times for each company that you're going through and you're likely to go through at least a thousand. That means you're literally looking at around 8,000 numbers, which would be quicker if you simply used a clustering technique to auto sort them.

As A Day Trader

The difference between penny trading and day trading is a little bit different in how you go about it, but also very much the same. You are still looking at variables such as debt and you are looking at debt history, but you are also including the link to your history that you would not normally include with penny trading. This is due to the fact that as a day trader, you are not looking at just companies that are trying to survive but companies that are making a decent profit. Therefore, you are looking for very much the same variables but for a different reason.

This means that you might have to tweak your algorithms slightly so that you include extra variables in your classifications. You primarily want to look at the companies that are making a decent amount of money, have a decent margin of safety, and have a considerable P/E ratio. Once you develop a machine learning algorithm that takes these into account, you can then further develop it so that it takes other variables into account and thus becomes more precise. However, this is how you might create the machine learning algorithm to make very quick assumptions as to companies that are worth your time within minutes of you giving it data. In my honest opinion, I would probably have three different machine learning algorithms running at different times because this allows different variations of knowledge to be coming through. I would have your standard Financial machine learning algorithm running and using all the variables I would normally use to judge a company, a machine learning algorithm that just determines whether the company is surviving or thriving versus dying, and a final algorithm that determines trending companies at that time. These will give you, mostly, different results but the sweet spots are the companies that fit into 2 categories, primarily trending with the first or second ML algorithm.

Clustering Techniques

Partitioning is your Friend

I've already shown you how to choose the best times but we both know that the best way to determine a company is if it shows similar features. Therefore, we look at the PE ratio, debt, total revenue, and many other distinctive features. Now, if you are searching through thousands if not millions of companies, it's going to be difficult to look at each and every variable. Let's say that you were only looking at five variables for the company, you would then need to look at 5 million variables if you went with millions of companies and 5000 variables if you were dealing with thousands of companies. Lastly, you would then also need to take note of which companies were good and which companies were bad. Therefore, you may actually accidentally check the same company twice and waste time doing that.

The best way to figure out which company is worth time investment is really to utilize the same variables that you would look at and run a partitioning algorithm on them. Specifically, on each of the variables you would look at, you would run a partitioning algorithm. Therefore, you would look at a clustering algorithm that organizes all of the PE ratios according to randomize spots so that you have high PE ratios, medium PE ratios, and low PE ratios. You could do this with total revenue, debt, and pretty much any of the other variables. However, they are not going to be your final conclusion. Since you compare all of the numbers together, you then take the companies that are put into the categories that you like the most and do the comparison that way. What it does is it shrinks down the number of companies that there are to a group of companies that you can make assessments of based off of comparisons.

1. Cluster like-variable companies

2. Extract companies with Preferred variables

3. Use standard comparative statistics on those companies to determine the value

K-Means Clustering

K-means clustering is actually a really simple, reiterative technique based on mean calculations. For every category of data you want to measure, you place a randomized "centroid". You then calculate the distance from each datapoint to each centroid and create a relation to the nearest centroid for that datapoint. Once this is done for all centroids, you calculate for the mean of each cluster. Using the mean of each cluster, we recluster using the mean values to measure the distance rather than the randomized value we started out with. Finally, if that doesn't change the clustering, then we calculate and keep track of the variation between the clusters. It will repeat this process as many times as you want, but only return the one with the best variation.

The primary problem with this algorithm is that it's sensitive to outlier data. Therefore, if you have companies that have huge amounts of debt, they are likely to throw off your results. If you really want this to work, you have to run through this algorithm, see what it produces, and if you notice that you are getting skewed results, you need to find the companies with the huge amounts of debt and remove them. So, if you notice that a company with $60k of debt is in the same group as the company with $900k of debt, you know that the second company is an outlier. You can learn more about K-means clustering here: https://www.datascience.com/blog/k-means-clustering .

K-medoids

This algorithm does very much the same thing as the previous clustering algorithm, however, it is less sensitive to outliers because of the medoidshift algorithm. The only problem and the main reason why this algorithm is not very popular is due to the computational requirements that it needs to do. Therefore, if you are iterating over hundreds, thousands, or even millions of companies, it's going to take you significantly longer to make the necessary calculations to get to the end result when compared to the first algorithm that I listed here.

You can find out more about K-medoids here: http://www.math.le.ac.uk/people/ag153/homepage/KmeansKmedoids/Kmeans_Kmedoids.html .

CLARANS

Due to the lack of optimization of the k-medoids algorithm, there have been many redesigns for how the algorithm goes about doing the same calculation. This algorithm stands for Clustering Large Applications with Randomized Search. This means that it is specifically designed to run a very similar version of the previous algorithm, but it's also changed so that it is better for large-scale applications.

Best software to automate your trading decisions

A Quick Notice

Much of the applications that I list here I will say that you can get a lot of information from. You have to realize before you try to go get that information that I am talking about information interception, which is a lot different than just accessing information. Information interception is where you figure out how the information is being aggregated into a program and then you intercept that information for your own uses. It's not necessarily hacking, but a lot of people would view it in the same light. Now, there are some software here that does not require that you do information interception but most of the software are all-in-one platforms where the user is not expected to need anything else outside of the platform. Therefore, the digital infrastructure is not necessarily there for some of us who want the information these programs are delivering.

Fully Automated Online Trading: Wealth Simple

Wealthsimple is by far the best in its' category because it *literally* handles everything, but it does come at a cost of 0.05% annually as of right now. The reason why it is the best is that it determines how it will trade based on your current life choices. Therefore, the older you are the less risk is associated with the profile. If you are single, it places you at higher risk but if you have a lot of debt it puts you at lower risk. There's a medium-sized list of pre-applied options that will determine your risks. Therefore, it is choosing how your account will trade based on your current lifestyle. The only way it can determine which options are of the lowest risk is either by using Machine Learning algorithms or quarterly predictions reports.

You can then determine what you want your profile to trade into. For instance, if you are an Islamic individual, there are specific ways in which it is tradition to trade as an Islamist. It has an option to trade like this that also cooperates with your current risk level. If you want to invest in the "green" solutions like Tesla and the like, there's also a plan for that. If you just want it to make more money regardless, there's also that option. It not only determines how risky your trades will be, but you can also let it handle the social responsibility of your profile.

You can put as much as you want into the platform, which means you can start out with $1. It's able to do this by investing in partial shares, which means there's no limit in how much you can invest. However, when you do invest more, you earn more over time. Once you reach over a certain limit, you can then utilize your invested membership to take advantage of certain benefits.

Thinkorswim Ameritrade Practice Software

Starting off with Real Money is a bad idea in most cases if you are new to trading. Additionally, if your machine learning algorithm is new to this game as well, you don't want it handling real money until you know it's good enough to not lose you a ton of money. On top of all of that, it doesn't make financial sense to create your own financial trading software to simulate a stock market just to then make a machine learning algorithm to game it. You run into issues such as trying to make it for the ML algorithm, which means it'll be easier or too hard for it. Generally, you want a software that's meant to train humans how to trade stocks.

This is where Thinkorswim comes into play as it is a significantly complex system that simulates stock trading. In addition to this, it also includes a Paper Money mode, which allows you to trade with Fake Money. This makes it the perfect software to base the construction of the machine learning algorithm off of, provide datasets to the algorithm, and ultimately optimize the algorithm for performance.

It is important for the construction of the machine learning algorithm because of the real-life comparison the software can have. In addition to this, the software is completely biased towards helping stock traders make more money. This means that instead of simply handling a stock market chart, the software provides ups and downs, opens and closes, and many more variables to work with. It's all in one spot, which means that you can siphon the information into the algorithm.

Lastly, the software does not take into account that a machine learning algorithm might be used in conjunction with the software. This means it is neutral for your neural network, which is a best-case scenario.

TradeStation Automated Software

This piece of software is a bit on the pricey side, so you might want to make sure that you have some Deep Pockets before you try to use this software. They do have a simulated service, which means that it is excellent for those who might want to use it to train their models. However, if you go with the service you will need to meet a monthly minimum activity to ensure that your account doesn't become "dead" as they really only want active traders on their platform. This software has a lot of bells and whistles that will make your trading a lot easier if you decide to utilize it instead of having a machine learning algorithm do all of what you need by itself. They have excellent fees whenever you're trying to trade, some of the lowest in the industry and even better for machine learning purposes, they also come with free real-time data on trading if you're into options trading. They've been an outstanding company for years and they have a lot of history, going back all the way to 1982 where they are listed as the first online trading platform. They're not the best but they are really good at providing you with the services you need to get some automated features going. Additionally, they have an extensive macro library and come with a mobile option.

Etna Automatic Trading Software

This is another software that does a really good job at providing a simulator and a mobile option for traders. I included this software on the list primarily because of its customizability and developer friendliness. Unlike many of the options here, you will actually find that this software company advertises more to developers than traders. They have several APIs and web trading capabilities rather than solely being based on desktop technology. This means that if you plan to make your automated trading as a platform, you can actually develop and deploy using their technology without having to create your own brokerage software.

eSignal Automated Trading Software

This software is something that I would call data puke. I don't really use the term very often, but I find that it exquisitely describes how much information comes from this program. You get access to real-time data, like much of the software here, but on top of that, you actually get access to real-time news concerning trade information and a litany of course videos and graphing abilities that I find lacking in many of the other software. While data puke does grant that you will be getting a lot of information, the actual application can be a little bit intimidating once you open it up. This is because almost all of the information is presented to you outright, but just like Gimp, if you spend some time with it, it can actually be a very good software. Likewise, because it is a data puke software, you will have plenty to feed your machine language algorithm if you decide to create one.

Tensorflow

Honestly, if you're really wanting to, you can build your own, but this is going to take quite a bit of time. The software that I've mentioned here are special privilege software and there's a reason why they almost always take a certain percentage of the money. You see, the market is a game of a fee on top of a fee on top of a fee and the pattern continues until it eventually hits you. This means that you are essentially paying the highest fee possible when you go with services like this because they are able to give you the access... at a fee knowing you likely can't go to the market floor. A lot of money is made by providing these platforms and really all they are doing is just providing you with access and information. The more information they can provide you, the more likely you are going to be willing to pay for their platform. This means they have an incentive to pump their software as full of data as possible, which means that a Legion of Developers is really what you're trying to weigh against you developing it on your own. They have experience with the actual software, they have been doing it for numerous years usually, and they know exactly how to incentivize people. Sure, you can develop your own, but it's like trying to make an AAA game by yourself with next to no money to hire anyone else. You can do it, it's just gonna take a really long time.

However, if you plan to make a machine learning algorithm that utilizes their software, then I would highly suggest going with Tensorflow. You see, there are tons of tutorials that will tell you how to make a machine learning algorithm that fits your purposes. Specifically, tutorials on YouTube because tutorials written on a website are usually bookmarked and forgotten. So, you can utilize most of the software here to collect the information that you need and then you can develop a machine learning algorithm that sends input to the application you are aggregating information from. Tensorflow is the easiest way to go about this because they are giving you access to their neural networks and all you have to do is implement the programming logic to take advantage of it. Once you feed the information into the machine, you can actually create an interface that encapsulates the software that you're taking information from and then send the inputs that software is expecting as a click for a buy and a trade. It's up to you really, but it's going to take longer if you try to make the platform yourself and then try to make the machine learning algorithm.

Conclusion

Machine Learning Can Be a Revolutionary Change in Your Life

Machine learning has helped many industries do many things so it's no surprise that in an industry full of numbers, machine learning seems to be on the rise. Machine learning does have a lot of capabilities ahead of it and it has proven to be somewhat successful for many individuals. However, something odd comes up when many people think of the stock market as an unpredictable beast yet we are designing machines to study and sell or buy based off of what they study. This means that the stock market can actually be predictable but within what margin?

Machine learning algorithms are very efficient at what they do. The mathematics needed to understand them literally comes from almost all walks of arithmetic and other libraries of mathematics. The formula that I introduced before had a little bit of linear algebra, a little bit of calculus, and a little bit of geometry in it. It was a huge and scary

equation for some yet enlightening for many others.

I would have to say that if you can get machine learning

algorithm to work for you, it will be revolutionary to how you handle

stocks. In fact, if machine-learning proves to be extremely successful, I

think that the average person will begin to engage themselves more

with the stock market. A huge barrier with the stock market is simply

understanding how it works and it is not easy to find resources that are

willing to divulge this information to you. I find this to be ironic

because the more people that are in the stock market, the more money

that can be made. If the stock market only had one person in it, it

wouldn't make any money. However, because millions of people around

the world are playing the stock market game, billions to trillions are

traded annually.

In addition to this, Stock Market trading may actually just

become as commonplace as a checking or savings account. This would

be where a person simply puts a certain amount of money in their stocking account and lets the machine make the money. This means that everyone who actively works as a stock trader will essentially no longer need to spend hours and weeks of their time studying the market because we have machines to do that. They will still make money, it's just that many won't find why spending so much time handling the math themselves would benefit them.

Final Thoughts

As I already said, machine learning has many benefits to it and it could be a great opportunity. However, machine-learning is still very much an experimental type of programming. You shouldn't 100% rely on machine learning to make all of your decisions for you because there are huge risks to allowing that to happen. It will take time to train a machine learning algorithm to trade like you trade, but if you give it enough effort, you may find that machine trading is a lot more efficient than trying to trade on your own.

MACHINE LEARNING IN PYTHON

Hands on Machine Learning with Python Tools, Concepts and Techniques

Disclaimer

Table of Contents

What is Machine Learning?

How Programming Normally Works

The usual method of programming is quite linear, even in places where it seems nonlinear. The most common "insult" that some programmers use to refer to machine learning is that it is just a bunch of if... else statements where the machine is not actually learning. It is very easy to understand how these programmers come to understand this, but it is important to realize that they are only half right.

Let's look at how something like a website and Photoshop works, considering how widely the manner in which they operate is different. A website is a collection of HTML, CSS, and Javascript with whatever backend code implementation they plan to use. The website itself does not normally install anything on the user desktop and utilizes features that are already there.

The only mechanism that provides change is the web browser itself and it is only when the web browser supports changes in those languages do those languages *really* have access to new features. In order to construct the front-end of the website, one has to load the HTML, which will then load the CSS in the Head or the Body areas of the page and load the Javascript in, usually, the Body area of the page near the footer. Therefore, it is linearly loaded no matter how interconnected the web pages may seem.

In Photoshop, the implementation is definitely different due to the fact that it is a program that must be installed on a computer. To the average individual, Photoshop looks like a self-contained unit that can be utilized on every platform. However, Photoshop must utilize and have access to graphical standards only found in drivers for Graphics Cards. In order to draw a line, Photoshop normally has to make a call to the Direct X 11 or Direct X 12 or Vulkan or OpenGL libraries. No one really knows which library it calls to or if it calls to all of them, but all graphics-based programs have to call on existing libraries. This doesn't become apparent until the program encounters an error.

You might ask how I know this and it really has to deal with the variation of Graphics Cards on the market. You have Intel, AMD, and NVidia all making their own versions of Graphics Chips, with each version of these chips running on the previously mentioned libraries and even older ones. With AMD alone, I know that the past 10 years have seen Direct X 9, Direct X 10, Direct X 11, and Vulkan chip libraries. These libraries provide a consistent basis for function calls across the variety of Graphics Chips in the market. It would be impractical for Adobe, developers of Photoshop, to create their software from complete scratch for each Graphics Chip in existence when there are pre-existing libraries that other companies maintain that cut the workload significantly.

Therefore, in order for a program like Photoshop to even work, it has to have a linear access to already implemented resources. Photoshop, itself, is very modular but still linear. You can see this in how it structures its' menus. I click on Filter to find the Blur category where I can use the Gaussian Blur equation. Photoshop can be seen more like a library of different image related equations that have sub-equations to ultimately create a linear stack of *Layers* as they are referred to in Photoshop. Therefore, while the tools are modular, they are nested linearly and applied to the image in a chronologically linear methodology.

Having this in mind and having seen programs and websites work like this for decades, it is understandable that Machine Learning could be seen as nothing more than if… else statements. The problem doesn't rely on how programming works, but rather on how if… else statements are seen. For instance, *if true then this else then that* is a valid way to teach new programmers how to understand if… else statements. The programmers who compare Machine Learning to this could say *if (feature has curve) then feature is a, b, c… else feature is L, A, E…* and this could very well be a valid representation of how a network might work. However, that is how the human mind works and we learn all the time so what's the problem?

How We Define Learning

The problem, therefore, is the definition of what it means to learn, and this is indeed a philosophical discussion. You might have been asking why I have laid this out in such a manner, but it is truly important to understand that machine learning works differently than the average programming as it has been practiced. It is different not because of *how* it is programmed, but with what *intent* it is programmed for. This is why the philosophy is also important as it determines how one goes about making and implementing machine learning.

How does the human mind learn? It learns through practicing until it gets *it* mostly right. Therefore, our recognition usually fails us the first few times that we attempt to apply it. It is only through repeated failure that *human minds* find their Gradient Descent. Gradient Descent is how Machine Learning works, but exactly what is it? It is a mathematical equation given to us by Calculus and while it has many applications, Machine Learning uses it to measure the amount of error an algorithm has and move towards less error.

The philosophy behind Machine Learning is to define how *human minds* would normally classify Features mathematically at their most fundamental levels. Once we have this definition, we then begin to write algorithms that are designed to find these features in a more general sense because we humans don't make things with perfection as it would be in a computer world. For instance, while a circle is a circle in the human world it will eventually boil down into a line if we zoom in close enough. The computational world views it from a mathematical equation, which means it will never become a line no matter how much we zoom into it. Once we have written the drafts of these programs, these *if... else statements*, we begin to repeatedly test them to see how accurate they are when applied. This produces an error rate with each test and the goal is to make the error rate drop via a Gradient Descent.

In Calculus, this Gradient Descent is really just an X and Y plot line that curves with hills and valleys. The goal for those developing the Machine Learning algorithms is to create an error rate that exists inside of the lowest possible valley. Each error rate represents a plot point. However, this is still very much a Linear Program because we make it, test it, and change it to make it *more* correct and this doesn't constitute as learning. Learning requires that an algorithm is able to review past mistakes, use those mistakes to get better results, and make fewer mistakes. Thus, the key to unlocking a Learning Algorithm is how one can make the algorithm *remember and change* its' algorithm for better results.

The Cleverness of Recursive Programming

When looking at Machine Learning programs, the most common theme you will see is that those programmers will often run these programs thousands of times to see what it does. While we could very well explain the different methodologies behind how one goes about teaching an algorithm, the most important facet is that the programmer is looking for the correct weights and biases to get the best Gradient Descent. Here, I am going to discuss one of the many types of algorithms used that will help you understand why most programmers make their programs recursive.

Let us say that we have a dataset of 100 randomized characters and we want our algorithm to recognize letters. The first method is where we have Supervised Learning, which is where we know the correct answer for every character that goes into our Machine Learning algorithm. The goal here would be to feed the character through the algorithm, see if it guessed correctly, and change values if it got it wrong. We could do this by hand, but this is usually time-consuming and human error prone, like reusing values by accident. When it comes to individual feature detection, this is manageable. You may only have to test 100 times for each feature to make sure it detects the fundamentals.

However, when you have to detect if the Machine Learning algorithm can utilize those feature detection nodes in unison, it becomes a mathematical nightmare to do it by hand. Basically, you can think of it as a factorial equation with each feature detection node adding one more to the factorial. Therefore, if you are testing for seven features, you would need to test it by hand seven factorial or 5,040 times. Instead, we would be much better off if we had the program detect when it was wrong, have it change its' own values, and then reattempt to guess correctly. This, by definition, is a recursive algorithm, which is the most common way Machine Learning is practiced. However, having a known database with known values is still Supervised Learning regardless of whether it is recursive or not, recursion just makes the process faster.

The importance of recursion inside of Machine Learning cannot be understated because this is how the algorithm teaches itself from then on. Imagine having to correct every error Google Voice Recognition provided by hand; it's simply impossible for one human. Thus, recursion allows the programmer to run test batches via Supervised or Unsupervised and glean information on whether it is working well or something needs to be changed in the fundamentals. While I may not have defined all of the points, this is generally how Machine Learning works and is applied.

The Core of ML is Feature Detection

Now, I have talked a lot about Feature Detection without actually defining what it is and this is because it is an abstract concept rather than a defined item. For instance, when you look at the letter A, it will have different features to it than the letter a. Instead of looking for a direct definition, you would look for features like a straight line or a curve in the letter to determine parts of a whole. Parts of a whole is a great way to think about how Feature Detection works because that is how *all* Machine Learning algorithms sift through the data.

In order to create a feature, you have to Classify or Categorize those features that you define. The entirety of creating features is ultimately to determine "what does it mean?" because what does it mean if the letter has a curve? The natural answer would mean that it could only be part of a smaller set of structures. Let us go through the process of *detecting* or *recognizing* "Creek". Right off the bat we have something unique; the "C" is bigger than the other letters, which means it is capitalized. We could run a matrix to determine the size of all the individual characters. This would take the total amount of letters it could be down to 26, effectively cutting the search size down by half. Then we could notice that it has a curve, which would likely cut it down to just 10 of those letters. Now we could notice it is an open circle, which further cuts it down to G and C. Lastly, we could notice that it doesn't have a line in the middle, which gives us a guess of C. We would then follow a similar process for each of the characters in the string.

The important part to notice is that the Features we created were from noticeable differences in the data. Additionally, each "notice" was applicable to every letter we tried to recognize, which meant we could reuse those feature detection tools. This is how you create Features from what seems like random data no matter what algorithm you might be using at the time, but you really only have to start out with one feature and then it becomes much easier to define even more features.

The Development of "Neural Nodes"

As I have related to in the past, we in Machine Learning often reference our ideas about Machine Learning from how our *human mind* works. This is because the mind is the one reference point we can most relate to and the scientific basis for which we already have a significant scientific background (Psychology, Anatomy, ect..). This allows us to theorize how we might make a machine capable of the same *useful* mechanisms we have as humans.

The research essentially boiled down to what we know as the Perceptron, which is the first type of neural network ever to be conceived of and implemented successfully.

```
function node(x){
    if(weight * x + bias > 0){
        return 1;
    }else{
        return 0;
    }
}
```

The Perceptron is binary classifier that uses the code above as a basis of its' equation (yes, the real version is a bit more complicated) and it is the simplest example of what we call a neural node. Now, for the longest time, the weights and bias referred in this algorithm constantly confused me as to what they associate themselves to. Here is the equation for Slope Intercept Form:

$$y = mx + b$$

In Slope Intercept Form, we are trying to find where a singular point intercepts the slope of a line. In a Binary Classifier, we are attempting to classify if a singular point is before or after a slope. In the case of the program we utilized, we are simply detecting if the classified value is above or below zero. You probably already noticed how the Slope Intercept Form is very similar to that of the Binary classifier, which was how I actually understood weights and biases for the first time.

In Slope Intercept Form, the m is where we obtain our slope, the b is our Y intercept, and our x is just our x plot point. Therefore, our weight is how we define what the slope is and then the bias is really our Y intercept. Instead of equalizing it to Y, we simply determine if the result is below Y or above Y. Normally, the weights are actually randomized because all we have is data and we're trying to find the correlation between 0 and 1. The value of b is determined by how far off the slope is supposed to be from 0. Needless to say, that I am simplifying this far further than most academic papers might, but this is how it works in the most basic sense.

Once we have our neuron up and running, we now need to make it into a node and this is what we need the Activation Function for. There are several different Activation Functions out there and our basis of the Perceptron does have its' own activation by either returning a 1 or a 0 otherwise known as Binary Activation. The Activation Function is used to determine if it is going to send a value to the next node in the network or not, which actually makes itself a neural node rather than a neuron.

Finally, what is Machine Learning?

While I might have delved a little bit into the Perceptron, it is only a single example in the library of different neurons utilized in a neural network. In this chapter, we have covered the misguided notion that machine learning is just a bunch of if… else statements and we have also gone over what actually makes up a neural network, but now it is time to finally answer the question of "What is Machine Learning?" and there's a couple answers to the question depending on how you approach it.

The definition of machine learning is the ability for a machine to make decisions based on data and then based on the outcomes of those decisions, change how it works so that it is capable of making better decisions. For instance, let's go through a logic tree.

- Is this a Box? -

  -> Does box have vertical lines?

- Yes

  -> Does box have horizontal lines?

- Yes

  -> Does each horizontal line vector with a vertical line?

- Yes

-> Box prediction

- Yes

Answer = No

In such a situation, the questions asked of the neural network about an item may suggest that it is a box, but it could also just be four corners that don't interconnect with each corner. The answers would be right, but there's no room for self-correction and the logic tree is incomplete. This would be a machine learning algorithm if we simply added better questions, which is why neural networks are usually quite large.

The other definition of machine learning is that it is the ability for a machine to remember incorrect predictions and correct predictions, which through the comparison of those memories that the machine makes more correct predictions. However, you run into the definition problem with this and so the application of machine learning algorithm becomes societally based. For instance, how do we determine if another human is an enemy? If you answered that they must have caused us harm, then we could say every human is an enemy because they take usable air away from us so they can live.

The last definition and perhaps the more preferred is that machine learning is an algorithm designed to provide estimated predictions based on mathematical classifications and datasets. This allows machine learning to be an abstract construct that is defined by the application it is used in. These are the primary answers to a seemingly simple yet complex question, which really represents the potential complexity and difficulty of machine learning.

Why use Python?

Python is Quick to Pick Up

Python is an extremely easy language to pick up and was originally designed to be easy to pick up. The knowledge of how Python works was originally based on work the creator had performed in ABC with interpreted languages. Very much like how many of the languages that were developed at the time, Guido van Rossum had issues with how the languages he worked on actually worked. One Christmas, he decided to write Python and he so named it after Monty Python's Flying Circus.

Alright, so it wasn't specifically built for ease of use, which is Ruby's claim to fame. Instead, the developer of the language saw the need to have an easy interface to low level items. However, historically, Python is usually the introductory programming languages for those not going into Computer Science as a way to convince them that programming is not difficult.

Python is an interpreted language, which means that you don't need to waste time making definitions for items that should be quite obvious. With languages like C++ and Java, the benefit to having control over the definitions primarily has to deal with memory management and speed. Since Python is an interpreted language, many see a very drastic decrease in speed, but most don't notice that there are other arguments in play. Most Machine Learning is *done* with Python, but when it is put into production the odds of the program being translated to a language like C, like C++, or like Java are very high. This is because Python is very good at allowing developers to refine the mechanics without having to wield a foray of definition errors off with a cyber stick.

Not only does this cut down on time but it also allows users to pick up the language very quickly. Those familiar with programming and diving into Machine Learning can utilize background knowledge to easily jump into Python without much trouble. Most anyone who has a background in data analysis or manipulation can jump into Python because of how important Matlab is to the data analytical world. Matlab is an all-encompassing mathematical library that helps those manipulating data to create graphs and since Python can manipulate Excel, the other Math Lord software, Python is kind of a programming math king language. Due to the nomenclature of Python, those sufficiently studied in mathematics will easily transition into programming because almost all of the syntax is math-based. Pretty much, if you're doing something with math and it involves programming, Python is your heavyweight champion.

This implies that Python is really easy to pick up if you're naturally inclined to be a mathematician regardless of level, which is primarily true of most programming. However, it's also not just the mathematical side that benefits Python but also Blender, The Sims 4 and Eve Online. These are very well-known programs that harness the power of Python, but Blender in particular does a really good job because a sufficient enough Blender user can use Python in Blender.

You might be wondering why I bring up these three, specific software applications of Python to the forefront.

Blender is a 3D Graphics Modelling and Animation software, which is vastly popular due to the fact that it has been free and will seemingly remain free. There are hundreds of thousands of tutorials on YouTube and ever more on the internet about how to use Blender. With the widespread popularity of the program, it is only natural that when you become an advanced user of Blender that you begin to use the Python console. By using this console, it is like unlocking an entirely different version of Blender where a lot more can be done.

As for Eve Online and The Sims 4, these are massive video games that have culturally affected every country they have been in. Video games, in my opinion, is a huge reason why younger generations show interest in learning how to program. In the beginning adventures of many programmers, a primary question is "what the most popular video game language is" so that those programmers can learn the language to create games. Those who are a fan of Civilizations and Eve

Online tend to profess Python because it is used extensively between the two. Eve Online, though, is popular because of its' cultural impact and so is The Sims 4.

In Eve Online, it has been debated as whether the crime committed in the game constitutes as actual crime in the real world. This is because there have been several controversial incidents where users scam, pirate, and induce corporate espionage in the virtual game. The developers purposely refuse to do anything about it and enforce the idea that players should not sink money into the game for items if they're not prepared to lose it. The organization controlling the game actually created an internal affairs segment to ensure the developers did not take advantage of their position, back in a time when computer-based organizations hadn't commonly thought this to be a standard practice. There are wars inside cyberspace that result in huge "wartime" losses in a way never seen before. There was a 6,142-player war that eventually came to be known as the Siege of 9-4 that cost over one million dollars in real currency.

There is a subculture in Eve Online that produces mods to allow the game to be operated and to operate in a much more organized manner. These mods have to be written in the same language as the original product in order to avoid compatibility issues both online and offline. While the User Interface is offline, much of the data that is utilized in Eve Online mods rely on what is coming in through the connection, so it needs to handle specific Python data typing.

YouTube is the primary reason why The Sims 4 became so popular because it was a center of attention for many massive YouTubers such as DanTDM, Pewdiepie, and others. It was part of the meme culture and unlike Blender or Eve Online, it didn't hold much importance beyond that. With older games though, the code is usually much easier to break because people found vulnerabilities in it. Additionally, it also helps if that same code either has poor encryption or no encryption, which is where the modding community for Python comes in again. As I explained with Eve Online, in order to mod a game and prevent compatibility issues, the mod usually uses the same language that was used to make the game. With over 500 mods in the

Sims 4 community, a game that is quite old by our standards has a very active modding community.

This means that a lot of modders for these two games and more wind up using Python as a result. Therefore, subcultures of the genre would naturally be swayed to hold Python above other languages. Add in the likelihood that these subcultures could pick up on Python quickly and you have a fanbase spawned entirely from other areas of the economic tree for entirely different reasons. This leads into the next reason why Machine Learning is Python-Popular.

Python has a Massive Science and Community-Based Culture

You might be asking why I chose these three specific software products to explain why Python is quick to pick up because you do not see a correlation between those products and machine learning. They do have a correlation and that is in the amount of data that's utilized. For Blender, this community focuses on graphical standards and making much of what we see in games, animated movies, and the list goes on. There's a massive amount of geometrical and geospatial data in

Blender. Eve Online is an economic system with a massive player base powering it, which means that there's a ton of data passing through the mods and being displayed for players. Sims 4 has to deal with an emotional setting where player interactions with non-player entities result in emotional and financial consequences in the game. All three products utilize and harness massive quantities.

However, at the same time that they utilize massive quantities they are also very simplistic in the needs that are met. For instance, in Blender you may just want to automate a specific skeleton that would be a slight variation of the original. A mutant may have six or four fingers, teeth, a few teeth or all teeth, and at this point you can think of this like a game like Spore. Instead of spending hundreds of hours trying to create all the different variations, one can simply write a clever Python script that will take the base model and modify it. In order to do this, a person has to be a competent person in several difficult mathematical fields.

In the Eve Online modding community, some of the more basic mods are simply creating User Interface mechanisms that list team player statistics, player locations, and even financial charts like one would see in a Wall Street office. This level of mathematics is more about knowing Statistics 101 and beyond with a little bit of Predictive Analysis to help control the flows of the market. When everybody is trying to gain with the same math, it becomes a mathematical war among players.

In Sims 4, most of the mods are either visual or psychological, which requires an understanding of two different fields. Those that are visual not only require the knowledge of reverse engineering but also the same level of mathematics needed in Blender Python. Those that are psychological have to know Community Economics, Ecosystem Analysis, and other fields in order to make a mod that doesn't break the game by making it impossible to win or taking away from the purpose of the main game.

Python is a language that sort of specializes in all of these types of mathematics. While the main language has a heavy amount of support for code libraries like OpenCV, Matlab's, PyChart, and similar mathematically based code libraries, the modding communities provide a different set of resources. Machine learning or neural networks has really been focused on how to utilize machines to better benefit people and companies. Text-based recognition networks are designed to take literary texts and translate them into digital words much faster than humans can. Predictive networks have been helping to create safer trafficking systems on the streets, safer security measures in highly sensitive spots, and even help try and game the economic gambling market. Image recognition has been tested for recognizing medical diseases better and faster than doctors, recognizing faces in crowded zones, and another level of conspiratorial purposes as shown by Hollywood. Any mathematics using image recognition needs to rely on the same mathematical skills Blender programmers use. Predictive networks utilize the same types of mathematics in Eve Online and Sims 4. Therefore, you can begin to understand why Machine Learning has begun to take its' place in video games as well, often modifying or replacing existing AI to help challenge players into being better.

Those who study and implement machine learning are usually academic scholars, governmental bodies, corporations, and hobbyists. These are the four primary categories that make up those that are interested in large coding investments like Machine Learning. Most of the information you're going to find is a cross between academic scholars and hobbyists, so it's important to understand where the bulk of your information is coming from. You're not going to get the algorithm that government bodies or corporations use because that's how they make money and control. The academic scholars may have been the ones that created the technology, but the hobbyists made it prosper. It was not the motherboard kit maker that posted forum questions or made StackOverflow, but the hobbyists that used them and shared their issues. The number one thing any programmer will find themselves doing on (practically) a daily basis is looking up a potential solution to an issue they have. These hobbyists created the code libraries and some of the algorithms you'll be using while you study this material. They are the ones who decided how languages work and where they're applied, which is why they often make it easy to understand how they made it. This is the essence of why Python is so easy to pick up. Every one of the groups prefer to get things done in the now rather than later, which is why Python has the syntax that it does.

Python's Syntax Creates Faster Turnaround Times

The art of writing is the most difficult part of novel writing and programming. Not only does the author need to carefully pick the best words to use for the task they're trying to get done, but the language also needs to support those words. Everyone also knows that time is money, so when you combine the two concepts together you get the need to express your ideas quickly to make the most money. This is the primary reason why interpreted languages like JavaScript, Python, PHP, and Ruby exist.

As a result of this, when you are attempting to put an idea into practice then the faster you can create the structure for that idea the faster you can initiate that idea. Since Python users want to have a quick solution, they rely on Python syntax to take care of the nitty, gritty

details while they get their idea sorted out. In Python, definition of a

function takes this:

```
def myFunction():
```

Meanwhile, in C++

```
int main(){ return 0; }
```

As you can see, Python does not require that you predict what

type of return result that you're going to get while C++ does. This is

very important because you may not always know what you want in the

idea that you're trying to prosper. Are you going to want an array of

coordinates, an integer, or a string of text that your singular function

recognized? You haven't even started writing the application and you

have to predict what the result will be at the end, which doesn't make a

lot of sense. This is why Python syntax is much easier to pick up

because the language does not assume you know what you want so it provides a lot more freedom. However, that does not mean that your program is going to stay in the same language.

Python is the Development Language, C is the Production Language

One of the most common practices in programming is the art of translating one program into another program for an advantage. Python syntax may make creating the idea quick, but Python is considerably slow because the program has to interpret that program before it can run that program. As a result, many companies and organizations often take the original program and translate it to C or "bridge it" to C. This takes the step of interpreting the program out of the equation, which drastically reduces the amount of time it takes to execute the program.

As a result, the translated program is usually also the final step in the entire process of making a program. The problem then becomes an issue of whether the functions need to be updated or not. When you work on a program like SolidWorks or Microsoft Word, you make a single program that's going to run the same way unless it is a different version. You can make software that doesn't need constant maintenance, but those ways are actually dying very slowly as many people conform to the idea of the IoT.

The Internet of Things is the idea that all usable items in your house will eventually be connected to the internet. Therefore, imagine that your car registered you are headed home, checks your schedule and sees that you're not going anywhere else. The screen on the dashboard asks if you are headed home and if you want dinner. You then press the "Yes" button on the screen that then tells your fridge, oven, and microwave combo unit to begin making dinner. It then uses the approved container that allows it to remove the lid and puts it into the oven because it's a 30 min dish versus a 30-minute driving route. You walk up to your door, which automatically opens up because it recognizes your IoT key and then closes behind you as it registers you entering and leaving its' scanning region. You are now home with a fresh cooked meal waiting to be pulled out of the oven that you don't need to turn off because the sensors inside the oven tell it to cook it no more as it is now edible.

The problem with the IoT form of life is that it requires the internet and when things are on the internet, actively on the internet like Facebook or StackOverflow, they are almost always never in production. You might find this weird as a concept, but these websites are always developing new features and updating older features. This means, essentially, that they are never truly a final product but rather a series of final products, which would be better known as continued development. So, why is Python important when it comes to an environment of continued development and machine learning?

Python is Web Compatible

Python is often considered to be a "back-end" language. C++ is not considered to be a "back-end" language and C is definitely not a web language. While Python can perform regular software actions, it also has the ability to host web pages and this means that a person can take their machine learning software online. This also means that if your website runs on a different language, you can use what's known as a Python wrapper to create a bridge between your language and the machine learning you're attempting to associate it with, but this is not

233

ideal. You have data transfer rates, Python interpretation, and the program to execute so the cost of running a system like that is unacceptable.

Instead, the most common solution is to just have a Python server that takes the information in and executes the functions like a regular program. This allows your website to function as it normally does and creates an interface where Python can collect the information, send it to the C++ program, and then Python sends it back. It is a necessary cost in most cases because there's not a lot for Python to do other than receive, send, receive back, and then send back. Most of the time is taken up by C++ being executed and the bandwidth involved, which means you don't have to wait long on Python interpretation.

Therefore, if you have a toaster that learns how toasted someone wants their toast to be then you can make a website that the user can control this and other appliances. You can then use this data to make agriculture predictions or sell the information.

So, let's recap, Python is used most often in Machine Learning because:

- It was already popular in Mathematics, Gaming, and Graphic Design culture.

- The syntax makes it fast to develop and prototype.

- The syntax is extremely abstract, which allows an individual to program without needing to predict the result or define everything.

- Due to its' cultural impact, Python has coding libraries and forums that are their own ecosystems, creating massive troves of research and error correction information. This natural backing makes it easier for new programmers to learn the language.

- Finally, Python is already a web capable language, which means that most neural networks (Facial Recognition, Economic Trade Analysis and Prediction, ect.) can easily gather information from the web through Python. Additionally, since a bridge can be made between Python and C++ for a speed advantage, Python is often a first choice for any combination involving analytics and web technologies.

Regression Analysis using Python

What is Regression Analysis?

Now, if you recall, I actually started this book off by comparing the y-intercept formula and Regression Analysis actually utilizes an equation very similar to this.

$$y = \beta_0 + \beta_1 x + \epsilon$$

In this equation, we are saying that the y is equal to that of the y-intercept population parameter plus the slope population parameter plus the error term. The error term is labeled as such because it represents for the unexplained variation in the equation for solving **y**. Essentially, this is the part of the equation we're ultimately trying to reduce so that we can have accurate results. Ultimately, the ideal equation is found below:

$$E(y) = \beta_0 + \beta_1 x$$

The **expected y,** which can also be donated as $\hat{y}$ when working with sample data, is equal to that of the sum of the y-intercept population and slope population parameters. Now, let's back up here because you might not actually understand how to calculate the error term in this equation when you first start out.

Alright, so let's start out by going to the market to negotiate how many trinkets we can buy and for how much. There is no set price for the trinkets and we are buying the same trinket from different sellers. Here is a table that lists 10 sellers and their different price points for selling their trinket to us.

Sellers	Price Points
1	5
2	17
3	11
4	8
5	14
6	20
7	2
8	8
9	11
10	10

In this table, we can see that we have a price point for each of them. Now, can you predict what the next price point will be for the 11th seller in such a graph? Since the only definition we have right now is the price point, the best next prediction will be the mean. The mean is the amount at which there is a 50/50 expectation that it will be the correct prediction. To calculate the mean, we simply add up all the price points and divide them by the number of price points that there are. The mean for this table is 10.6.

Now that we have out mean value, we can now calculate for our residuals. A residual is a number that deviates from the mean value. While technically all of the values deviate from this, the residual is how much it deviates from the mean. Thus, here is our new table.

Sellers	Price Points	Residuals
1	5	-5.6
2	17	6.4
3	11	0.4
4	8	-2.6
5	14	3.4
6	20	9.4
7	2	-8.6
8	8	-2.6
9	11	0.4
10	10	-0.6

Something to keep in mind here is that the Residuals are the actual Errors we are talking about in Regression Analysis. Now we need to find the **Sum of Squared Errors** or the Sum of Squared Residuals. In the following table, I have done just that.

Residuals	Squared Resid	SSE
-5.6	31.36	260.4
6.4	40.96	
0.4	0.16	
-2.6	6.76	
3.4	11.56	
9.4	88.36	
-8.6	73.96	
-2.6	6.76	
0.4	0.16	
-0.6	0.36	

Now, you might be wondering why we are going through these very specific steps. Simple Linear Regression, the most basic form of Regression Analysis, is based on reducing the SSE (Sum of Squared Errors) to create a **Best Fit Line**. In Linear Regression, we are comparing this SSE that we got when we assumed there was only 1-set of categorizing data (the dependent variable) to another that has 2 sets of categorizing data (the independent and dependent variable). Linear Regression is a part of a special type of mathematics known as

Bivariate Statistics. Bivariate means that there are two variables or variations in the Statistics that you may be studying.

In Linear Regression, the Y-Axis is meant to be the "Why?" while the X-Axis is meant to be the "Explanation" of the data. Therefore, "Why is price point 1 at 5?" and then our X-Axis would be used to explain why it is there. You may have also seen that I included a slope in a previous table and that is because the data does have a slope. However, when you are using an equation like this:

$$\hat{y} = b_0 + b_1 x$$

Where the sample data expected y is equal to that of the sample data y-intercept population plus the slope population in order to calculate for 1 variable, you are using a slope of zero. Now we're going to go ahead and add a 2nd variable to our equation, which will be how much money it costs the seller to actually buy the trinket from the person who made it. We will call this new variable the **Initial Cost** as it is the initial cost of the trinket before it is marked up for profit by other sellers. The Price Point is DEPENDENT on the Initial Cost, which a very important distinction to make. Remember that I said that the Y-Axis is the why, thus the Price Point is our new Y plot point and the Initial Cost is our X plot point. Now, here comes a *new* equation:

$$min \sum (y_i - \hat{y}_i)^2$$

This is known as the **Least Squares** equation. If you remember correctly, the **hat of y** is the result we get from our sample data where our Initial Cost didn't exist. The reason why the second y in this equation does not have a hat is because this y is what we will observe of the actual data. The hat of y is our predicted data while the regular y is that actual data. We will be finding the difference of these two, but not on a graph by graph basis. This equation requires us to minimize the sum of the squared differences of each predicted y with each observed y in a linear progression. Here is the new data we will be utilizing:

Sellers	Price Points	Initial Cost
1	5	2.5
2	17	8.5
3	11	5.5
4	8	4
5	14	7
6	20	10
7	2	1
8	8	4
9	11	5.5
10	10	5

I am aware of how clear cut this is, but this is because we're utilizing fake data to make this easier to understand. In the real world, you could spend weeks only to find there's no correlation so for teaching purposes it is much better to have a mock scenario. Instead of looking at this data, you would be putting it in a Scatter Plot like this:

Price Points and Initial Cost

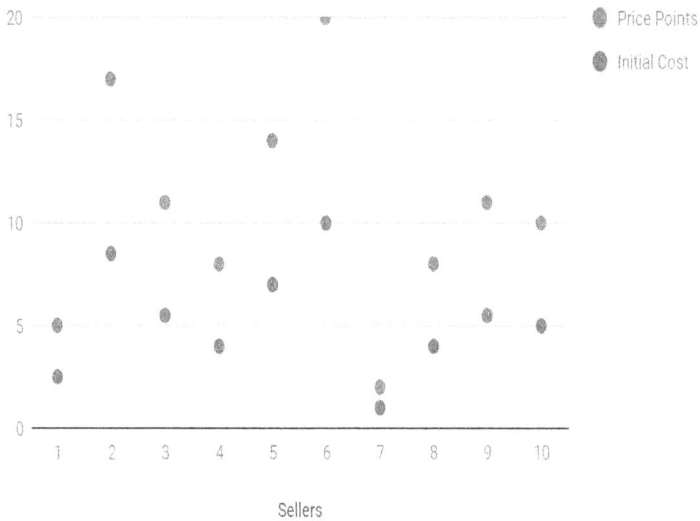

In this graph, it is not as clear that there is a correlation, and this is why data representation is key to seeing the relation between the two. For instance, if I ran this in a Line Graph, the correlation would be glaring:

Price Points and Initial Cost

The next step in this process is to find what is known as the **Centroid** and it represents the point at which our Regression Best Fit Line will pass through.

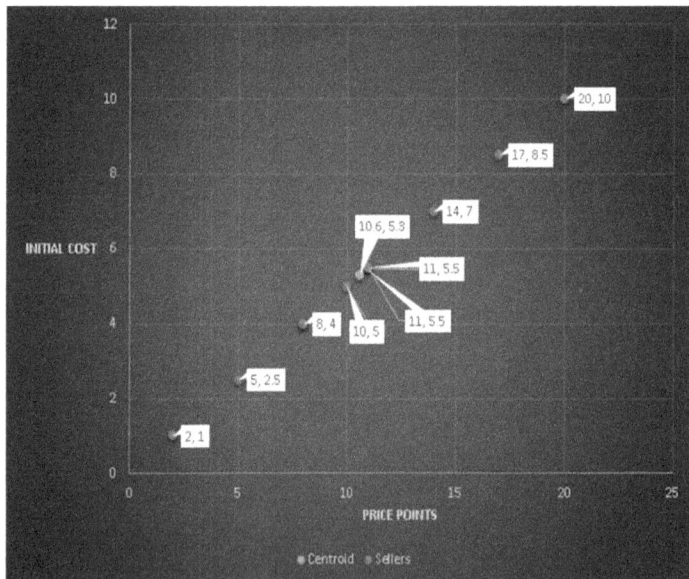

As you can already tell, our Regression Line will most likely go straight through that line. So, the first step in plotting this Regression Line is to find the Slope or the b_1 of our equation and that equation requires a bigger equation:

$$\frac{\sum(x_i - \bar{x})(y_i - \bar{y})}{\sum(x_i - \bar{x})^2}$$

Now, this could very well look rather scary at first, but this equation is actually quite simple. On top, we are finding the difference between the independent (initial cost) variable as x and the mean of that independent variable as well as between the dependent variable as y and the mean of the dependent variable. Then we multiply those together. Once we are all done doing these to all of the variables, we then add all of those results together before dividing. On the bottom of our division, we take the independent variable and subtract the mean of the independent variable, but then we square that result. Once we do this to all of them, we add all the results together. Here it is in Python. I prefer to view math in code quite often rather than the equation:

247

```python
independent_var =
[2.5,8.5,5.5,4,7,10,1,4,5.5,5]
independent_mean = 5.3
dependent_var = [5,17,11,8,14,20,2,8,11,10]
dependent_mean = 10.6
def slope(independent_var, independent_mean,
dependent_var, dependent_mean):
    d = []
    x = []
    top = 0
    y = 0
    for i in range(len(independent_var)):
        x.append(independent_var[i] -
independent_mean)
    for i in range(len(dependent_var)):
        d.append(dependent_var[i] -
dependent_mean)
    for i in range(len(x)):
        top += x[i] * d[i]
    for i in range(len(independent_var)):
        y += (independent_var[i] -
independent_mean)**2
    print(top/y)
    return top/y
pass
```

As you can see, it's relatively basic in what needs to be done but

you now need to feed it into the other side of the equation.

$$b_0 = \bar{y} - b_1\bar{x}$$

The answer to our slope was 2. This will now be multiplied

against the independent mean (x bar) and subtracted from the dependent

mean (y bar) to equal b_0. In our case, using the modified algorithm:

```
independent_var =
[2.5,8.5,5.5,4,7,10,1,4,5.5,5]
independent_mean = 5.3
dependent_var =
[5,17,11,8,14,20,2,8,11,10]
dependent_mean = 10.6
def slope(independent_var,
independent_mean, dependent_var,
dependent_mean):
        d = []
        x = []
        top = 0
        y = 0
        for i in
range(len(independent_var)):
                x.append(independent_var[i]
- independent_mean)
        for i in
range(len(dependent_var)):
                d.append(dependent_var[i] -
dependent_mean)
        for i in range(len(x)):
                top += x[i] * d[i]
        for i in
```

```
range(len(independent_var)):
        y += (independent_var[i] -
independent_mean)**2
    print(top/y)
    return top/y
pass
def y_intercept(independent_mean,
dependent_mean, slope):
    return (dependent_mean - (slope *
independent_mean))
    pass
print(y_intercept(independent_mean,depe
ndent_mean,slope(independent_var,indepe
ndent_mean,dependent_var,dependent_mean
)))
```

This means that the value is 0, which is not surprising in our

case but now we've got to put this back into slope intercept form. Here,

that would be:

$$\hat{y}_i = 0 + 2x$$

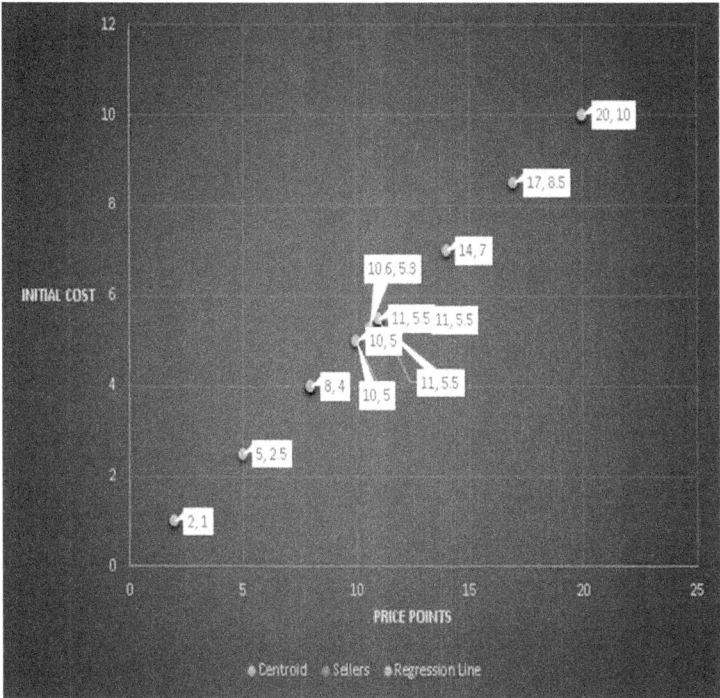

As you can see from this graph, the Regression Line is a completely perfect match, which means we have done the right calculations.

Why use Regression Analysis?

There Are Two Primary Reasons: Reason 1 Correlation and Importance

Regression analysis allows us to see if two variables have a correlation with each other if we decide to study those two variables. For instance, let's say that the variable that we used to determine our price points was based off of how much it cost the seller to buy it from the manufacturer. If the seller or sellers rather had a consistent method of charging a certain amount of money above others, it will show clearly within our graph. Remember that regression analysis is really the study of two different variables to see if that those variables have a correlation. Therefore, the variable that determines the price point of the seller will rise equally for the amount of money the seller bought the material or trinket from the manufacturer. The beautiful part about regression analysis is that once we put it into a graph, it becomes conceptually clear to us that there is a correlation by drawing a statistically significant line that represents the rate of change. If one can draw a line almost perfectly through the two data points and there is no sum of squared errors, then there is a high likelihood that there is a correlation between the two variables.

When we are dealing with massive amounts of information, we have to find a way to determine what information is important and what information is just a waste of our time. Let us say that we have a spreadsheet that tells us the day the trinket was sold on, the hour that the trinket was sold in, the city the trinket was sold in, the name of the person who sold the trinket, the street that trinket was bought on, the building number that the street that the trinket was bought on was in relation to, and the locations of where the materials for the trinket was sourced from. This spreadsheet is an absolute nightmare in terms of how much data you have to sit through because what if you have over a thousand different sellers on this list and there doesn't seem to be any correlation between them?

Regression analysis allows you to take each category and compare it to each other. In the list of items that are in our data sheet, we would have 7 different sets of data we would need to compare. Of those seven, it is most likely that we could create an association between where it was sold and where it was made. These would be negligible findings, but they would still have an association. With regression analysis, we are able to determine whether it is an important variable that we need to research. For instance, yes, it is likely that all of the trinkets were made in the same location. This doesn't really have a great importance other than knowing where it came from. Likewise, there's also, probably, not much significance to the name of the person who sold it since we are buying it from several different sellers, the building number that's near the place you bought it from, and a few others. What's more important is likely to be in association with the day that it was sold on, the street that the trinket was sold on, and maybe where the materials were sourced from. The reason why I say that these might be important is because we know that there are certain days of the week where items are on sale versus items that are sold at their normal price. We know that it is more likely that a seller will place a deal on an item that doesn't sell very well on their slowest day of the week. We also know that if the resources are more expensive to source from that this will ultimately lead to a higher price. These have reasons behind them that can be further explored, whereas the other ones don't have a direct correlation and thus don't have a direct importance to the exploratory search that is regression analysis.

Reason 2: Supervised Learning

The other reason to use regression analysis is because regression analysis is how you perform supervised learning. When you are testing the variables inside of a neural network, you are performing a regression analysis at every layer of your network. Let us look at how we might Define a line in a picture. Theoretically, you could Define a line as saying that in order for it to be a vertical line you must have a significant contrast difference between the previous pixel value, the current pixel value, and the next pixel value. You could create a matrix that looks like this:

$$\begin{bmatrix} 000 \\ 010 \\ 010 \\ 010 \end{bmatrix}$$

This is the beginning of a pattern recognition program, another form of a neural network. In this neural network you are comparing two values at a time. You compare the previous pixel value with the current pixel value and then you compare the current pixel value with the next pixel value. This ultimately determines whether there is a significant importance to the pixel you are currently on and this is how pattern recognition programs work, except that they have different equations and different matrices to handle different patterns. When you have to compare two variables at a time, you will be utilizing linear regression or regression analysis. I say linear regression or regression analysis because there is more than one type of regression analysis, linear regression is just the most common. Linear regression refers to the fact that you can plot a linear line between the differences.

Clustering Analysis using Python

What is Clustering Analysis?

There is no one-way of doing a cluster analysis and it is more of a concept than a specific type of algorithm. I suppose the best way to describe the difference is to utilize two examples. Let's say that our first example is that of cat and dogs. How might you go about classifying these two animals? Well, you know that you'll generally look for differences. Now, be careful because **cluster analysis is abstractly defined**. This means that you cannot check for specific features, you are just looking for things that are true of all living things. Therefore, you might look for size, width, and other physical features. Let's say you created a cluster analysis that could tell the difference between cats and dogs.

Now, could you see the same success if you applied this algorithm to trying to cluster server motherboards and personal computer motherboards? Honestly, the most you would get is an almost fair success rate of size. This is because motherboards like the Raspberry Pi can actually be a server motherboard. You would have to develop a different algorithm based on different **abstract** parameters. Cluster Analysis requires that you not know ahead of time what features you are specifically looking for. You have to rely on trying to find any differences at all without knowing which difference will make the most impact because you don't know what features your area of interest will have.

How is Clustering Analysis different from previous Machine Learning algorithms?

Previously, Machine Learning algorithms had a primal understanding of what it needed to sift through. For cats and dogs, it knew that it would need to pay attention to the facial features as well as the shape of the animal itself. These features would be fine-tuned so that it would be much easier to pick up those features. The key aspect of these features is that they have finite parameters built on abstract concepts. Cluster Analysis allows a machine learning algorithm to have abstract parameters on abstract concepts.

Therefore, let's say that we're trying to find things that are worth looking into about a population. Instead of finding things to search for ourselves, we simply input what we know about a population and feed it to a cluster algorithm that attempts to sort and cluster relative data. The data that does cluster are the success zones of our algorithm and they become the new topic for further exploratory analysis.

A Theoretical Analysis of Clustering Analysis

Perhaps the most popular means of performing a Cluster Analysis is with the K means algorithm. Clustering, itself, is usually a part of the Discrete Unsupervised Learning algorithms group. This is how the K means Clustering algorithm works:

1. You make a best guess of how many K centroids you will need to make.

2. You systematically organize those K centroids to better align themselves with the clusters they are nearby finding the Euclidean distance.

260

3. Those new K centroids are now your labels for your clustered data.

So, since you basically know what a centroid is, we need to cover what Euclidean distance is, which is *this* calculation:

$$\sqrt{\sum_{i=1}^{n}(a_i - p_i)^2}$$

You might think that this equation looks blindingly complex, but it is important to note that this is simply calculating the Straight-Line distance between 2 points. To break it down into a more simplistic and less abrasive equation, this is what it looks like for 2 dimensions:

$$d = \sqrt{(x_2 - x_1)^2 + (y_2 - y_1)^2}$$

Or, alternatively,

$$d = \sqrt{(x_1 - x_2)^2 + (y_1 - y_2)^2}$$

As you can see, this form of the equation is far less scary and so long as you have 2 sets of coordinates, you can just plug in the numbers to get your result. The previous equation is for the Nth dimension, which means it calculates for more than 1 set and it is iterative. Literally, it translates to "calculate the sum of each of the indices, starting with one, as the subtraction of pair 1 index 1 and pair 2 index 1, squared and each consecutive pair. Then square root the entirety of the sum". Here is a Python implementation of the Euclidean distance.

```
import math
a_array = [(10, 1),(9,
6),(3,2),(7,4),(8,5)]
p_array =
[(8,3),(7,4),(5,9),(2,10),(6,1)]
def euclidean_distance(a_array,
p_array):
        result = 0
        for i in range(len(a_array)):
            x_1 = a_array[i][0]
            x_2 = p_array[i][0]
            y_1 = a_array[i][1]
            y_2 = p_array[i][1]
            result += (x_1 - x_2)**2 +
(y_1 - y_2)**2
        return math.sqrt(result)
print(euclidean_distance(a_array,p_arr
ay))
```

Working Clustering Analysis in a Real-World Application with

Results

Alright, so for this (and for licensing reasons) I have generated a

random spreadsheet with data. There are 1,000 companies with their

injury percentages, employee counts, and the percentages of whether

they have safety goggles and gloves or not. The goal of this cluster

analysis is to determine whether there is any relationship between these

numbers. Let's look at the Employee Count vs Injuries:

263

Employee Count vs. Injuries

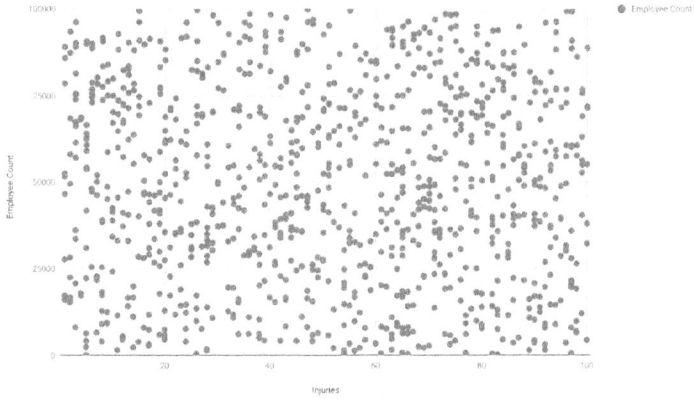

As we can see, it definitely doesn't look like it's associated based on raw numbers. If we look at Goggles and Gloves, we'll see the same thing.

Safety Goggles vs. Injuries

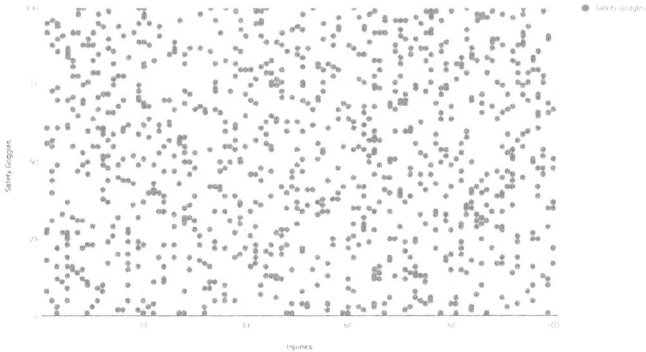

So, since we don't find an easy correlation there, let's abstract

the Percentage of Injuries vs Employee Count by studying the number

of

Gloves vs. Injuries

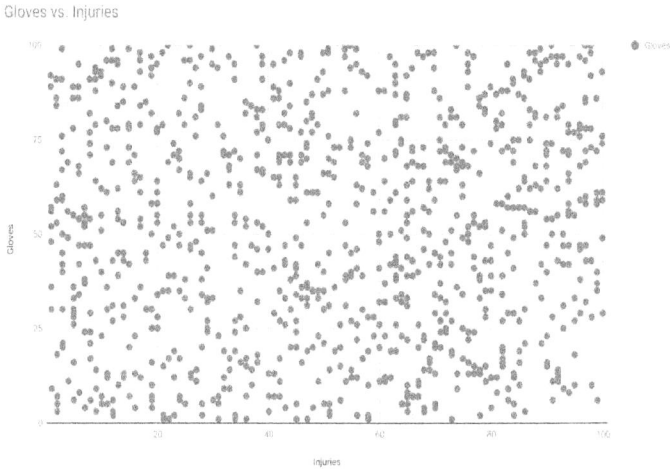

Employees injured vs Employees.

Employees Injured vs. Employee Count

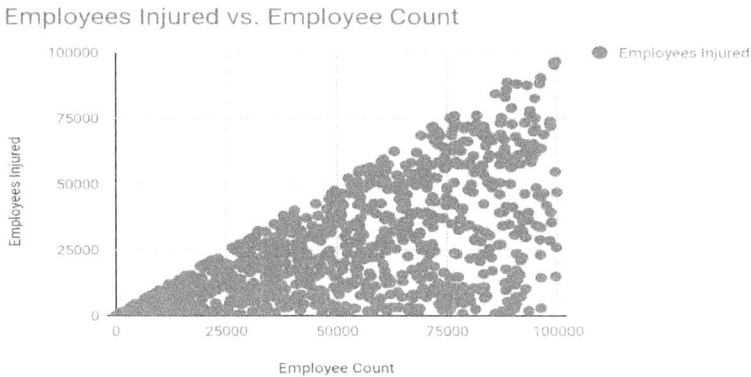

Wow! That's one heck of a difference that made. At this very

moment we've successfully performed a cluster analysis, without having to do complicated math, and can determine that for this data there is a strong correlation between how many employees one has and how many injuries they have. Remember that Cluster Analysis does not have one right way of doing it, so even if you are able to clearly tell from this graph and did not have to use an algorithm like K-means to get there, it still represents Cluster Analysis. However, in a normal practice, we would then try to see if there was an underlying association that causes this, but for the sake of this example and brevity, I believe this is sufficient.

Implementing an Artificial Neural Network

What are Artificial Neural Networks?

Artificial neural networks are a classification of algorithm mixed with conceptual analysis of how biology, psychology, and circuitry integration work.

What is Conceptual Analysis?

A lot of people equate conceptual analysis to being that of philosophizing about a concept. The problem is that philosophizing about a specific topic, while useful in its own right, doesn't really break down the aspects of how to associate such concepts of philosophy to real-world Applications. Philosophy is the art of getting to the root of something via a route of logical reasoning. This is in fact a very useful tool in the programming world, the problem is that it's not conceptual analysis.

The best way to describe conceptual analysis is if you were to take a beer opener that you bought at a local convenience store and used it to open a soda bottle, the old ones. In such a case, you have made an association between the old type of soda bottle and the current type of glass beer. The beer opener works for both of them, but as the name denotes it was specifically designed for only opening beer. What we have done here is we have made a conceptual analysis of the device and found a different use for it. We understand the beer opener as a device that lifts metal lids off of glass containers if that lid meets a certain shape. Therefore, while only intended for opening beer bottles, we have analyzed the soda bottle and found that we can use the beer opener to open that type of bottle. The definition is to take a concept, look at it carefully, and see if that concept can be applied elsewhere.

The Study of Biology

The concept of the computer and, in fact, neural networks actually comes from our days of biology. There are three levels of Sciences, which each science ultimately led to the next science. However, biology was the start of it all. Biology is the study of life and how life works. When scientists began studying the physical makeup of the brain, they eventually came to find out that the brain was capable of housing electricity. However, the concept of a computer was envisioned even before that. All you need to do in order to effectively envision a computer is to think of something that causes a chain reaction. Perhaps my favorite study of computer mechanisms has to deal with enormous Domino puzzles where they use the Domino's to ultimately calculate a binary equation.

The term computer was actually invented around the 1600s where a writer referred to an accountant or person who worked with numbers all day as a computer or a person who computes calculations. We have had computers for a long time if computers are, indeed, this definition. As humans, we have tried to take the repetitive mechanisms of our daily life out of our lives. The first repetitive mechanism we tried to remove was the necessity to do mathematics with nothing more than a pen and paper. Perhaps my favorite example of this is the famous abacus. We do this by studying our own lives and then inventing a machine or mechanism that can ultimately do those repetitive tasks. The more we increased our mathematical knowledge, which is what was required in order to build these machines, the more the concept of a computer came into existence.

The Study of Circuitry

The study of mathematics and biology, along with the constant need to make our daily lives better, eventually led to the creation of circuitry. Thomas Edison is often credited as the first individual to ever create a successful light bulb while Isaac Newton is famous for a Litany of mathematical achievements, for which one of those came the concept of Wi-Fi. However, electricity itself was a subject that was of great interest to many of the ancient civilizations. In fact, it's rumored that some Mediterranean cultures had a sort of magic that allowed them to rub material on cat fur that would attract bird feathers. We have had laws of magnetism for centuries now, but the study of electricity ultimately came from studying how objects in life interact with each other.

However, it wasn't until around 1791 to about the middle 1800s that we truly began investing electricity into our lives. It was in this area of time that the famous Volta battery and Faraday Motor was invented, which was also during the time that the study of electromagnetism became a real thing. Once we hit the 19th century, scientists were basically coming out of the woodwork on electrical theories, concepts, and inventions. These different scientists would eventually become the linchpin of how electricity works and then it was only a matter of time until we reached computers.

The Study of Psychology

Now, up until this point, I haven't really mentioned anything about neural networks according to the Sciences. However, it's around the psychology and conceptual analysis of the human mind that neural networks truly began to emerge. The purpose of psychology is the study of the psyche or the mind in most of our definitions. This meant that there was an entire science solely dedicated to figuring out how we figure things out. Because of our understanding of biology, we understood that there had to be a chain of events that led to where we

271

were. Because of our understanding of circuitry, we understood that the mechanism for delivering information inside of our bodies was usually done with electricity.

These two concepts had a profound impact when we started studying how the mind makes decisions. While the neuron of a human mind is somewhat more sophisticated, it was understood that every question only ever had two answers. In fact, you can think of complicated questions that have more than one answer as a question tree. The human neuron takes electricity inside and then makes a choice as to whether the electricity will leave its neuron to go down another path or if it just stays there. Therefore, the question will result in an activation or a no activation. Yes, every question really has only two answers. The understanding of this concept meant that if you could create a decision algorithm, that could run in parallel with other algorithms like this, and the algorithm had a choice to activate or not activate then you could conceptually create a human neuron. That is, ultimately, the idea of a neural network but because computers are not humans, there are some complications.

Backpropagation - Linchpin of The Neural Network

This leads us to backpropagation, which is the linchpin of neural networking in general. Backpropagation is really just a mathematical optimization, which is, essentially, what all of neural networking really is. It takes a mathematical solution and optimizes for a better result, which is probably why so many people find it confusing.

Let us start off by examining what our neural network is made up of:

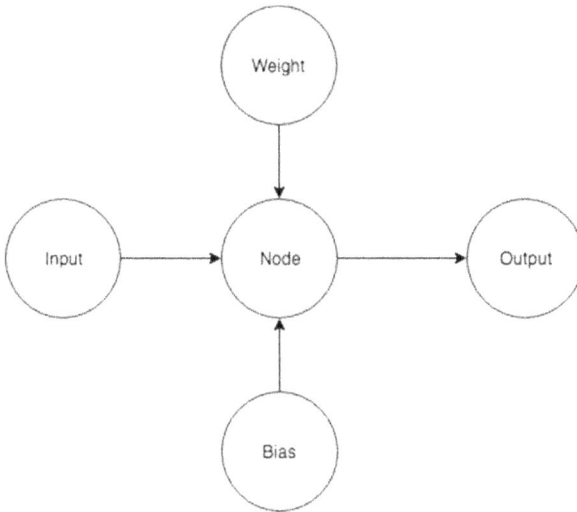

As you can see, there aren't many variables to take into account. In fact, it can really be summed up quite sequentially:

$$O = w + b + i$$

This makes for a rather simplistic looking equation, but now we need to add some complexity. Since we know that neural networks are made up of several nodes and not just one, let's go ahead and say we have 2 nodes. Well, now the equation has a cost to it. The reason why it has a cost to it is because Node 1 has to reach a certain value to send a value to Node 2 and Node 2 now has to combine its' value with Node 1 to reach a certain value to send its' value to Output. We will denote each Node as a and then each layer as a superscript L with a subtraction of 1 for each layer it is away from the output. Thus, our equation is now:

$$O = C = a^L + a^{L-n}$$

Therefore, the output is really the cost which is the combination of the last layer and all other layers a depends on (linearly so far) represented as n. Therefore, on the reverse side of this, we could actually say that our equation is this:

$$C(...) = (a^L - O)^2$$

Now, you might argue that this is a different equation and it definitely is because the output is now subtracted from what will ultimately be the value coming from the last layer. This is when you have a **desired** output and *not* when you just assign your output as the cost. Additionally, we have to make a change to our previous equation. Since we have more than one layer, we now have to add this in.

$$a^L = \sigma(w^L a^{L-1} + b^L)$$

So, let's breakdown what this is doing. Our very first node is a^{L-1}, which now affects the outcome of a^L. The value from that previous node is now multiplied against the weight value of our current node. Then we add the bias in before utilizing our **Activation Function** or, in this case, the sigmoid calculation. The Activation function ultimately determines if the value is significant enough to send a signal to the next node. At this point we have reached the calculation needed for a **Feedforward Network** or otherwise called a Neural Network by most. Feedforward means the information is being fed forward by the math.

Now we're gonna get into some Calculus because we have to talk about Derivatives. Essentially, we're trying to find out how much each variable the ultimate cost is *derived* from. So, here's a dump of math:

$$\frac{\partial C_0}{\partial w^L} = \frac{\partial x^L}{\partial w^L} \frac{\partial a^L}{\partial x^L} \frac{\partial C_0}{\partial a^L}$$

I know, I know. It's a lot to stuff down, so let's take small bites. What this is saying is that a variation in the Cost over a variation in the weight is equivalent to the variation in the previous layer output (denoted by x) over the current variation of weight, multiplied by the variation of the current layer value over the previous layer value variation, and finally multiplied by the current variant cost over the current layer value. I know, still a lot to take in but you can think of each one being based off of each other. You effect one set of ratios and all of the rest change with it and you change those ratios by either messing with the weights or biases, but this utilizes the weights ratio. To calculate for the biases, you simply switch the positioned weights with the values of the bias. This is known as the **Chain rule**. There is a bit more math to have a multi-layer multi-node network, but essentially you just now have to add indices to keep track of those values and add in the additional sums that come from multi-connected nodes. The understanding of these values and the manipulation of these values is *how* backpropagation works, which you then add these to the gradient descent concept I presented earlier. The reason why I am avoiding going all out is because, we would need to mathematically map something equivalent to this:

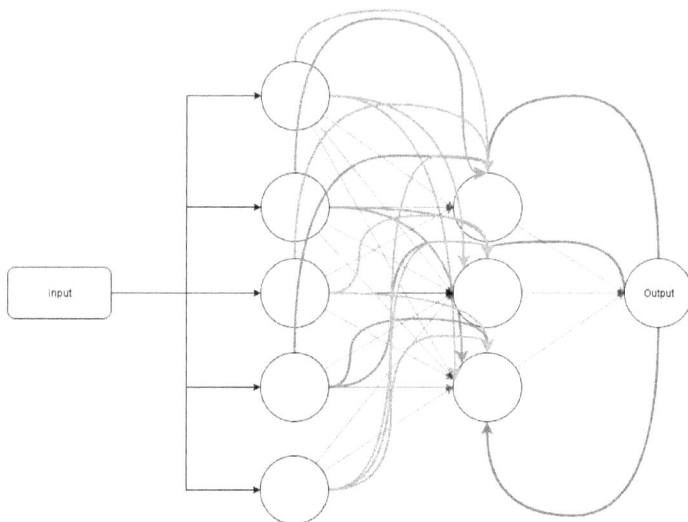

Take the principles you've just learned and add them to each colored line you see in this photograph, so you can see why understanding it is far more useful than walking through it. It's a very large concept and it will take you quite some time to fully digest it.

Backpropagating - An Applicable Program

So, for this, I'm going to bring in some fellow programming code as a real world applicable program for Backpropagation. The following code is underneath the https://opensource.org/licenses/MIT license and it is thanks to David Adler that we have this code. It was published May 30 of 2012 at ActiveState. It is open source code.

278

```python
import math
import random
import string

class NN:
  def __init__(self, NI, NH, NO):
    # number of nodes in layers
    self.ni = NI + 1 # +1 for bias
    self.nh = NH
    self.no = NO

    # initialize node-activations
    self.ai, self.ah, self.ao = [],[], []
    self.ai = [1.0]*self.ni
    self.ah = [1.0]*self.nh
    self.ao = [1.0]*self.no

    # create node weight matrices
    self.wi = makeMatrix (self.ni,
self.nh)
    self.wo = makeMatrix (self.nh,
self.no)
    # initialize node weights to random
vals
    randomizeMatrix ( self.wi, -0.2, 0.2 )
    randomizeMatrix ( self.wo, -2.0, 2.0 )
    # create last change in weights
matrices for momentum
    self.ci = makeMatrix (self.ni,
self.nh)
    self.co = makeMatrix (self.nh,
self.no)

  def runNN (self, inputs):
```

```
        if len(inputs) != self.ni-1:
        print 'incorrect number of inputs'

        for i in range(self.ni-1):
        self.ai[i] = inputs[i]

        for j in range(self.nh):
        sum = 0.0
        for i in range(self.ni):
        sum +=( self.ai[i] * self.wi[i][j]
)
        self.ah[j] = sigmoid (sum)

        for k in range(self.no):
        sum = 0.0
        for j in range(self.nh):
        sum +=( self.ah[j] * self.wo[j][k]
)
        self.ao[k] = sigmoid (sum)

        return self.ao


    def backPropagate (self, targets, N, M):
        #
http://www.youtube.com/watch?v=aVId8KMsdUU
&feature=BFa&list=LLldMCkmXl4j9_v0HeKdNcRA

        # calc output deltas
        # we want to find the instantaneous
rate of change of ( error with respect to
weight from node j to node k)
        # output_delta is defined as an
```

attribute of each ouput node. It is not
the final rate we need.
    # To get the final rate we must
multiply the delta by the activation of
the hidden layer node in question.
    # This multiplication is done
according to the chain rule as we are
taking the derivative of the activation
function
        # of the ouput node.
    # dE/dw[j][k] = (t[k] - ao[k]) * s'(
SUM(w[j][k]*ah[j])) * ah[j]
        output_deltas = [0.0] * self.no
        for k in range(self.no):
            error = targets[k] - self.ao[k]
            output_deltas[k] = error *
dsigmoid(self.ao[k])

        # update output weights
        for j in range(self.nh):
            for k in range(self.no):
                # output_deltas[k] * self.ah[j] is
the full derivative of
dError/dweight[j][k]
                change = output_deltas[k] *
self.ah[j]
                self.wo[j][k] += N*change +
M*self.co[j][k]
                self.co[j][k] = change

        # calc hidden deltas
        hidden_deltas = [0.0] * self.nh
        for j in range(self.nh):
            error = 0.0

281

```python
        for k in range(self.no):
            error += output_deltas[k] *
self.wo[j][k]
            hidden_deltas[j] = error *
dsigmoid(self.ah[j])

        #update input weights
        for i in range (self.ni):
            for j in range (self.nh):
                change = hidden_deltas[j] *
self.ai[i]
                #print
'activation',self.ai[i],'synapse',i,j,'cha
nge',change
                self.wi[i][j] += N*change +
M*self.ci[i][j]
                self.ci[i][j] = change

        # calc combined error
        # 1/2 for differential convenience &
**2 for modulus
        error = 0.0
        for k in range(len(targets)):
            error = 0.5 * (targets[k]-
self.ao[k])**2
        return error


    def weights(self):
        print 'Input weights:'
        for i in range(self.ni):
            print self.wi[i]
        print
        print 'Output weights:'
```

```python
        for j in range(self.nh):
            print self.wo[j]
        print ''

    def test(self, patterns):
        for p in patterns:
            inputs = p[0]
            print 'Inputs:', p[0], '-->',
self.runNN(inputs), '\tTarget', p[1]

    def train (self, patterns,
max_iterations = 1000, N=0.5, M=0.1):
        for i in range(max_iterations):
            for p in patterns:
                inputs = p[0]
                targets = p[1]
                self.runNN(inputs)
                error =
self.backPropagate(targets, N, M)
            if i % 50 == 0:
                print 'Combined error', error
        self.test(patterns)


def sigmoid (x):
    return math.tanh(x)

# the derivative of the sigmoid function
in terms of output
# proof here:
#
http://www.math10.com/en/algebra/hyperboli
c-functions/hyperbolic-functions.html
def dsigmoid (y):
```

283

```python
        return 1 - y**2

def makeMatrix ( I, J, fill=0.0):
    m = []
    for i in range(I):
        m.append([fill]*J)
    return m

def randomizeMatrix ( matrix, a, b):
    for i in range ( len (matrix) ):
        for j in range ( len (matrix[0]) ):
            matrix[i][j] = random.uniform(a,b)

def main ():
    pat = [
        [[0,0], [1]],
        [[0,1], [1]],
        [[1,0], [1]],
        [[1,1], [0]]
    ]
    myNN = NN ( 2, 2, 1)
    myNN.train(pat)
    if __name__ == "__main__":
        main()
```

How This Script and Others Like it Can be Useful

If you carefully look at the code, this book has a lot of what this

code talks about, more than just backpropagation. The constructor class

NN is actually making a Neural Node and several of them can make a

284

neural network. It also has the functionality to run that network and then backpropagate. This script is useful because it is a generalize script that allows you to use it in multiple scenarios and can be modified for your needs. Honestly, if you can fully understand *how* this script is working then you can understand *how* most neural networking and backpropagating scripts work.

The only con that I have with this script is that it needs to be modified to run on a Graphics card. This is great if you're running it on a PC without much information, but in order to get a power network you need a lot of cores. They are now selling "AI" specific Graphics cards as a Neural Network can only handle as much work as much it has access to processing cores. However, understanding this script with propel you quickly through your work on Machine Learning.

A 90 Day Plan for Machine Learning with Python

Concept

Day 0 - Day 7

The first seven days of this project should be dedicated to thinking about what you really want from your neural network. You should not spend these seven days just lounging about, but, rather, you should spend this time to discuss it with friends of the same mental capacity or colleagues to go over it as a concept. You should almost never make a blank statement of "I want to create [this] because of [this]" as it almost never works. This should really be a time where you take the time to refine the concept.

During the conversation, you should cover a few bases to ensure that this program is built and is built with the best purpose and moral intentions. When you build a neural network, you should question its' purpose. It might be fantastic to build a neural network that identifies the type of fecal matter something is, but what use does it have and where can you apply it? If it has a use, can you further refine that use so that it is specialized? Specializing reduces the amount of work that needs to be done in order to get a working product. The further you can specialize a neural network, the further that you can modulate this network. Modulation means networks can be developed in parallel, which means they can take a shorter amount of time to get done if the neural network is developed by a team.

Not only do you need to find out the specialization capabilities and what you can modulate, but you also need to know where to apply it and how to apply it. Let us take the algorithm that Google got in trouble with a few years back for as an example.

To put it as I understood what happened, Google created an algorithm that would determine the most associated searches towards a phrase. It was later found that if you made a search referencing a white man, job and work-related references were likely to show up. However, if you made the same search referencing a colored individual, you were more likely to see services where one could take a look at jail-time records and offenses. The reason why it ended up like this was because the network had found a correlation of women doing biased background checks on colored men versus white men. While there is a political stance here, there was a populated uproar against this result and the talk of racial bias in programming began. Therefore, when you go about creating a program that's allowed to "learn", you really need to also conceptualize where the program can go wrong.

During these discussions, you should really be recording them so that you can review them after this period.

Definition and Flow Charts

Day 8 - Day 14

Once you have spent quite some time on the concept, it's time to finally lay it down on paper. Provided you recorded the conversations you had, this step should be rather simplistic in nature. It's really important to define how the neural network is going to work and while you may have conceptualized and plugged up any holes during the concept phase, during the flowchart phase is when you begin to find how your particular neural network is going to operate. It's during this time that you really need to think about where the potential outliers that you don't want into the system will bake itself into the system.

Additionally, it is during this time that you can begin to conceptualize how long the neural network is going to take. Just to let you know, the more neural nodes that you can run, the less time it's going to take to ultimately train that Network. This means that you have to break the concepts down into their very, most basic elements in order for it to work at a fast and efficient rate. Flowcharts also help you determine which layers you're going to develop and in what order. Usually, most beginning networks start off with 1 to 2 hidden layers with an input and an output layer.

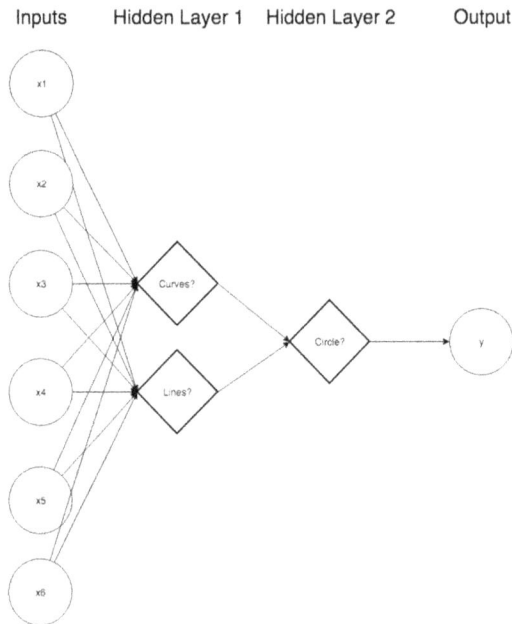

| Inputs | Hidden Layer 1 | Hidden Layer 2 | Output |

Hidden Layer 1 Development

Day 15 - Day 21

The hidden layer is where all of your neurons will ultimately be working, and it is named The Hidden Layer because you and your team are likely the only ones who will ever see it. A lot of people who use the hidden area will ultimately only ever see the inputs that they are using and seeing the result of the area that's hidden. It is during this time that you really need to try to get the most rudimentary algorithms down. During this phase, you should not be concerning yourself with

290

optimization or trying to correct the algorithm itself. Instead, you should be developing a very basic notion of how the algorithm is to work and then putting that into practice.

A lot of time is wasted in neural networks by trying to optimize your first go at a neural network or a specific type of neural network. There are a lot of ways to optimize algorithms so that they move faster, but that's not your goal in this first step. This first step should really be about trying to lay down the foundation of the beginnings of the first area of your neural network. Once you're able to test it and see that it works in a way that you kind of want it to work, then you can begin to make optimizations to make it faster and this is actually where correcting your neural network comes in.

Supervised Learning and Corrections

Day 22 - Day 28

During this particular set of days, you should be paying attention to the different weights and biases that your neural network is employing. It is also during this time that if you begin to see successful results, you now need to pay attention to how fast you are getting those results. Depending on the speed of the results, if you have any neural networks designed to rely on how fast a reaction happens, the speed of your algorithms could change the effect that the neural network has. Therefore, once you begin seeing that your neural network is working with a training set and doing very well, it's time to optimize the algorithms you're using.

The reason why I mention optimizing algorithms is because while doing individual tests may only take somewhere around a minute to two minutes, you have to expand that amount by the number of people going to be affected by this neural network. Therefore, if you have a billion people utilizing this neural network then you have to take into account that if it takes a minute or two to test this neural network then it is going to take a billion times a minute or two in order to test it on that scale. Once you have made some modifications to your algorithms so that they're faster and you are getting better results, it's now time to head into the next phase.

Unsupervised Learning and Corrections

Day 29 - Day 35

Now it's time to let your algorithm just run wild at this point and see what happens with it. There isn't really a way about going and doing a randomized test, but what most people do is they take enormous amounts of data from a company that may give it away freely or may sell it at a price and then do their tests on that data. The difference between supervised learning and unsupervised learning is that supervised learning is where you know what the answers will be for the output should your algorithm be designed correctly. Unsupervised learning is where you have no idea whether or not the output is correct without doing a manual calculation yourself as to what should happen. In fact, unsupervised learning is often utilized to find the necessary labels you might need for supervised learning.

Now, this is not to say that unsupervised learning is appropriate for all learning algorithms or neural networks. In fact, there are some cases where it makes sense to not do any unsupervised learning because you can always generate labels for the input that's coming into the neural network. In this case, you would simply just further optimize and test your neural network in much bigger batches. The idea of unsupervised learning is so that your algorithm is capable of defining its own labels for the inputs that it sees. The reason why there are so many neural networks out there is because there are just so many situations in which you might need one network over another. It really depends on what you're doing with your neural network.

Hidden Layer 2 Development

Day 36 - Day 42

Now it's time to take your current one hidden area network and expanded into further complexity. When you developed the flowchart, you should have gone from the most complex topics to the most basic topics to where one of two items would be the answer for the question. Now it is time to go up in complexity so that your neural network can become fully developed. Depending on the actual topic you chose to create a neural network, you may need more than two hidden layers in

294

order to actually do all of the necessary calculations. However, this would also expand this much further than 90 days and most people don't really need to utilize more than two layers.

The reason why most people don't need more than two layers is because what they are usually trying to do is a two-step process. You can often determine how many layers you're going to need by how many steps it takes to get to the output that you want. However, even if you do have more than one layer you tend to want to lower the number of layers that you might have because of the potential problem of the infinite gradient. The more layers that you have in a network, the more likely you're going to run up against the infinite gradient. Therefore, you should really spend about a week developing each additional layer and then spending about 2 weeks testing each layer so that they are compatible.

Supervised Learning and Corrections

Day 43 - Day 49

This time is going to be a little bit different than the last time that you tested your network because not only are you testing the first layer of your network but you're also testing the second layer of your network. It is important to realize this because your second layer is going to be based off of both how you created that layer but also how successful the previous layer is. This is why you need to ultimately keep track of the outputs of the first layer before they go into the second layer so that you can catch on to problems before they become unmanageable.

Additionally, once you begin seeing success in this area you also need to see if there are unnecessary nodes. One of the problems when developing a neural network is that your original conceptualization and flowchart design may take into account unnecessary elements. For instance, if I am trying to detect whether an object is a nose or not then I would likely want to detect whether there are holes such as a circle and whether there is a general shape of a nose. Instead of detecting for curvatures, which is detecting circles, I would really just need to text whether there are two Shadows that are parallel.

Detecting whether a pixel is darker than the previous pixel is far easier then detecting whether something curves. In such a case, it would actually be a optimization to change it to the simpler equation.

Unsupervised Learning and Corrections

Day 50 - Day 56

During this next phase, you should both be preparing to handle unknown data but also be exploring potential ways to reduce your neural network when possible. After this phase, you are likely to go through an Alpha Testing session where you unleash this neural network amongst the wild. Up until this point, almost all of the data that you have had needs to be static data. What I mean by Static data is even though you don't know what the data may result in, ultimately you know that there's only a specific amount of data that's going to be tested. When I refer to Dynamic data, I am referring to data that will either change while in use or data that has no seeming end to it. One of the problems that neural networking has to deal with is the sheer size of the tasks that are required to completely handle the problem the neural network was designed to handle. You can think of Google's voice typing as an example of such a situation. While they probably tested it on known datasets, the neural network now has to contend with an almost never-ending stream of data. Therefore, not only should you be determining whether your algorithm is working correctly but you should also be exploring the best avenues for deployment.

Now, you may have an economic prediction algorithm that is designed to try and figure out where the statistics of the stock market is going. Even in such a case where you would essentially only be running the machine on your computer, you also have to consider just how many businesses are on the stock market. Your test sample sizes might have only been utilizing 10 to 20 companies at a time because you only have one server core. This is the true test of optimization and a location of study for neural networks. You will likely need to create some sort of server Farm unless you are able to reduce the amount of calculations down to something that could be handled by a single server, but you are likely going to need at least a few graphics cards because each calculation has to be done by a core.

Alpha Testing

Day 57 - Day 63

Alpha Testing is a closed group testing and it really depends on what type of neural network you are developing as to whether you will essentially need Alpha Testing or not. Rather, it depends on what type of neural network you are developing that will determine whether you need people to be involved or if you can simply just let your algorithm into the wild.

The best way that I can describe Alpha Testing for new programmers is imagine if you were to make the most basic version of a product. You know that it's going to take at least half a year for you to make the full product that you want to make but you have no idea whether your idea for a product is going to be good or not. The truth of the matter is that the only way that you can check whether a product is going to be good or not is if you test it.

Let's discuss how big companies go about Alpha Testing their products because you and I are unknowingly apart of this. When Apple decided to begin developing facial recognition on their mobile devices, it became very apparent that Apple might have not done it in the most morally correct way that they could have. They claimed to have tested the technology against millions of faces but how did they get those millions of faces?

What most people don't know is that this facial recognition software is not a machine learning algorithm in the normal sense. Instead, what you have is a neural network that was developed to do one task really well that stays on the phone. This means that this neural network only gets upgrades when the entire phone actually does an upgrade. Unlike a normal neural network, which receives updates regularly over the internet, this is a neural network that works in a type of stasis where it has one specific track that stays common no matter what phone it is on. However, the implicit agreement occurred when you decided to buy a new phone. In order to get this new technology, you actually had to go buy a brand-new phone that was considered the flagship phone of that time. You became a Beta tester in this case and we have seen the results of this new technology. The Alpha Test was done within the company, a company that isn't heavy in diversity when it comes to their work staff. Therefore, you also heard issues with Americans that had a melatonin range not within Caucasian limits, which Apple blamed on deadlines. Alpha testing can propel or hinder your efforts depending on how you carry it out.

Feedback Reception and Corrections

Day 64 - Day 70

During this period, you should begin learning about the success rates of your neural network depending on how you Alpha tested it. It is during this time that you are either extending the amount of time you need to develop the neural that work, essentially going back to scratch and figuring out why your feedback was so negative, or you begin to head down the route of finding out how you're going to do Beta Testing. During an alpha test, your testers should be relatively close and there should generally be a non-disclosure agreement between who you're testing with. This is because you should only have a small population of people you should be testing at this point in time so that you can get a general idea of how successful it is on a much grander scale.

Additionally, this is the time that you need to make several different types of statistical analysis about how your neural network is working with unknown elements. This is because neural networks are never finished, and I don't really understand how companies can't seem to understand this. When you develop a neural network, you are developing a system that is always going to need change in order to become more accurate. You can't simply develop a neural network and just leave it be because it's as accurate as it is ever going to be. In fact, almost no website works like this. It is there in this stage that you begin to figure out how you're going to improve and expand on your neural network as a basic concept because once you develop a neural network, you have to begin finding out ways you can further apply that neural network. Mind you, this is tackling the subject from a business point of view. If you are just developing a neural network for fun, it's really just about the statistical analysis that will tell you where the weak parts of your neural network are. It's important to go through this step even if you're not going to deploy this in a business environment because it makes you more familiar with the process of doing this in a production environment.

Beta Testing

Day 71 - Day 77

During this period of time you should now be beta testing the technology, which means that you need to find a public way of releasing it and incentivizing complete strangers to download and test your technology. Most big companies offer incentives such as payment for people who beta test their technology because usually the beta testers are put into some type of situation where they have to work on a regular basis. The best-case example that I can think about explaining this with is the Uber driving experiment where the drivers were supposed to be Uber drivers but, at the same time, they mostly allowed the car to drive itself. This is a fairly new concept of self-driving cars that has come onto the market but has received serious backlash because of how dangerous this could be. Therefore, Uber has to pay somebody to be in the car and pay attention to what the car is doing so that the car doesn't actually put anyone at risk for their life. There were a few incidences where the driver actually fell asleep or wasn't paying attention and the car made a miscalculation or a mistake, which wound up with the injury of another as a result of it.

As a person who may just be creating a neural network to run in the wild, the best-case scenario I can think of where you would be able to test your neural network is to actually just share it with other machine learning enthusiasts. In fact, this would be a better route to take because they are far better at being able to provide you with critical feedback in what they experienced with your neural network. Most companies don't normally have this type of luxury though because when they did the Alpha Testing, they had the machine learning experts already critically analyze what it was doing and how it was doing it. It's a different situation for a hobbyist to do something like machine learning than a company to do it because the company usually has deep enough pockets to step things up much faster than the hobbyist. That doesn't make it impossible for the hobbyist, it just makes the process a little bit slower.

Feedback Reception and Corrections

Day 78 - Day 84

During this time, you should begin receiving feedback whether it is expertly put together or if it's just that your neural network didn't work for them. Feedback is useful no matter what level of grammatical correctness it comes from. Having had to sift through thousands of different reviews to see and gauge the feeling that I get from the public, all feedback will generally migrate towards key points that you should be paying attention to. The problem for many machine-learning enthusiasts in the very beginning is that they think that all feedback will either be good or bad. The truth of the matter is that feedback is a bit varied in the way that it's displayed, but the details are in the overall emotions that the reviewers reveal.

Perhaps the best way to describe it is an old type of doll that was sold before the 19th century. Let me tell you, those things are creepy. In fact, a lot of people agree with me that many of those types of dolls are creepy but the fact that they're creepy doesn't help. What does help is that they will often list the reason why it is creepy, which in this case is usually the eyes of the doll itself is the part that's creepy.

In a neural network, you will often see a habitual pattern as to what the neural network got wrong. Let us talk about facial recognition in such a mannerism. If you tested the facial recognition patterns and the overall feedback was that it didn't work in some cases, if you were given the ability to have feedback mechanisms such as the data that was associated with the face that the algorithm got wrong, you could see that it might be that you didn't account for nose variations. There are a lot of different noses out there and if you had followed the suggestion of just detecting two Shadows underneath the nose, there are noses that are shadowless because they are designed in a more crooked manner. It's just an evolutionary trait that you didn't take account of. It is during this time where you have randomized feedback that you're able to recognize the weak spots of your neural network and account for them before you go into production.

Production

Day 85 - Day 90

Now you should have reached the production cycle and the production cycle is basically where you find out where you can sell your product. This is more of the business side than anything to do with the machine learning site because at this point you now have a product and you are now spending your time finding a way to put that product into the hands of your customers.

Conclusion

Understanding Machine Learning

Unlike common misconceptions, machine learning, while it is a lot about a two-way decision, it is not simply about a bunch of if else statements. This concept is a very basic and generalized form of understanding what machine learning really is. However, in order to understand what machine learning is, you actually have to understand the intent of the program itself. The definition of insanity is to do the same thing over and over while expecting different results each time that you do it. The conception that machine learning is nothing more than a bunch of if else statements is the statement that machine learning does the same thing over and over. In fact, machine learning changes with each iteration that the if else statements run so while it is a bunch of if else statements, it is not linear programming. Instead, it is referred to as recursive programming. It's recursive because once it gets a result that is correct, it feeds that result back into its random set so that it can further along the gradient descent that you are trying to achieve.

This means that the programmers who assume that it is just a bunch of if else statements are partially correct but don't understand the underlying implications of what machine learning is. This is why it is very difficult to explain high-level concepts in very basic terms. It is a bunch of if else statements but with a different purpose than your average program.

This leads us to the next problem and that is that machine learning often has a specific form of algorithm and application. You will not utilize the same type of machine learning for population prediction or economic prediction that you might use for text-based prediction. This is because the algorithm has to take up data that comes in different forms and in different varieties, but more importantly is that the output is of a different intention. With population prediction and economic prediction, you are trying to find commonalities in between the data so that you can make better predictions as to how many people will be in a society in the future, which companies might do best during what periods, and essentially trying to predict what no normal human being could predict. On the other hand, a text-based prediction utilizes known mannerisms and parts of speech in order to determine what the next word of a sentence might be. These are two very different types of predictions, but one common thread that you will find between the two is that one set of predictions is based off of unpredictable data while the other set of predictions is based off of a predictable set of data.

This means that whenever you're developing a neural network, it's not just how the neural network has developed but with what intention that neural network will be developed. This is because neural networks are not generalized and almost all of them have a very specific purpose in mind. Unlike the human mind, neural networks are not designed to have staggering capabilities of adaptation. In fact, staggering capabilities of adaptation is not a humanoid trait because we adapt very slowly. It is common knowledge that you cannot spend a week trying to learn programming and understand all that there is about programming along with the best practices that come with programming. It takes a very long time to understand what is needed from you and how to go about doing it. Neural networks work in a very similar way and they are more linear than humans.

The Staggering Support of Python

A lot of people associate the success of python with the excessive Academia programming in general. I will have to say Python does a great job in the academic world where it applies to numbers and mathematical equations that need to be tested out very quickly. However, schools and places of learning are generally not the Pinnacle of what I see in terms of machine learning and applications for machine learning. In fact, most of the Python code in the world usually has very little to do with any Academia at all. The problem with computer science in general is that it has this image that it is only capable of existing because people go to school for it and you need to learn how to do things from school because you can't learn it in real life. There are two sides to this vision. You have the old school mentality that computers are an academic achievement and resource. You have the much newer version mentality that computers are primarily a source

311

where you go to learn pretty much anything that you want. The problem here is that there are more people in charge with the vision of the older school of mentality than there are of the newer mentality. Most of the people working in technology, in the highest regions of technology, usually did not have any science degree whatsoever because depending on how they came about with technology that degree might not have actually been a degree that existed.

Python is primarily popular because it's very easy to get a hold of, it's very forgiving in terms of the syntax that allows the programming to follow, and it's a rather old programming language compared to other programming languages in existence. Adding on top of this are the different communities that surround the language itself. As we mentioned, Blender is a graphics modeling system and since that covers everything from movies to video games to even commercials nowadays, that's a huge crowd. Then you have Eve online, which is a massive multiplayer video game that allows you to develop add-ons to the system that have to be in the same code that the system was developed in. Combine the modeling community of Eve online with the modeling community of the Sims 4 and you have a very large base of people who are interested in not just making things but also altering things. The difference between making something and altering something is that altering something in programming means that you have to test the code that's there repeatedly to figure out how the programmers programmed the code. The code is not all there at once and you have to reverse engineer how the programmers worked out how they were going to build the game. It's a very difficult, time-consuming process that many experts in the field do on a daily basis when it comes to Advanced military weapons systems and similar actions. That means that this entire Community is heavily reliant on analysis and Mathematics, which lead us into why Python is so popular amongst machine-learning enthusiasm.

Machine Language just takes advantage of what's already there.

A Set Track to Set Your Neural Network into the Wild

Concept

You should always begin your neural networks by determining exactly what concepts you want to tackle in your neural network. You can go out and make the best neural network you could ever make without actually knowing what it was intended for, but the problem is that in order to make the best neural network for a singular task is that you have to know what that single task is going to be. Einstein took part in developing one of the worst weapons in the world and it is one of the best forms of Destruction in the world, but they weren't necessarily focused on making it the most destructive thing in the world. In fact, many people of the Manhattan Project regret that they ever took part in the Manhattan Project because of how much destruction it caused. They didn't have a true intention on exactly what it was that they were creating but rather a general concept of trying to make a weapon that might change the war. Yes, they did make a weapon that changed the war, but it was also the scariest weapon we have ever made. This is the very value that I've tried to convey that you need to consider when you

314

make a neural network because while you may make a neural network, you don't always know exactly what that neural network is going to be used for unless you discuss it as a concept.

Design

Once you a fully figured out what your neural network is going to be as a concept, then it is time to lay down the foundations so that you can figure out how you can arrive at that concept in a morally obligated way. Your algorithm is not going to be the best if it prefers Caucasian faces over Asian faces or your algorithms not going to be the best if it only takes account for the profits of companies rather than the deficit of companies, which is actually where profit can be made. Additionally, this is where you begin laying out the infrastructure of how much you're going to need in order to get to the concept that you're trying to talk about or you're trying to design for. A lot of the problem with developing neural networks is finding out just how much resources you're going to need to make it work. Luckily, there are some Geniuses in the world that found very clever ways of developing a method of reducing the overall cost, but you still need to pay attention to how fast things are, how they will aggregate in much bigger quantities, and just how accurate your algorithms will be. This is not exactly the stage that you want to utilize to optimize these algorithms but rather make sure that your algorithm works in the first place and that you are able to scale that algorithm.

Develop

The development and testing phase are really one and the same, it's just that one happens before the other. In order to develop an algorithm, you have to do exploratory tests to find out what you're doing and then in order to improve upon an algorithm, you have to do tests. Therefore, in order to develop algorithms further you have to test further and so these two really go hand-in-hand. The difference being is that during the development phase you are either creating algorithms or you are optimizing out for them, which can often stand for correcting. You ideally want to be handling both at the same time and creating enough checkpoints so that you can see where you, yourself, made a diverging change in the algorithm that might have led to what's known as a domino consequence. You have immediate errors that are easily recognizable and usually changeable on the spot. Then you have what's known as a domino error, and the only way that you are able to find out what your domino error is if you are able to keep track of where you change things.

Test

There's a difference between a closed testing environment, an initial testing environment, and a beta testing environment. During the initial testing environment, your concern should primarily be whether you're getting the correct outputs, whether your algorithms are working at a decent rate, and whether you have any areas that you can improve upon. The way that you can test whether your algorithms are working at a decent rate is both linear and parallel as well as serial. Linear testing simply means that each and every algorithm fits a Big O notation equation that is satisfactory. Big O notation simply refers to how fast an algorithm will work over a given amount of time. Parallel equation vs. serial equation is very different. In a parallel equation, you essentially want all the results to come out at about the same amount of time. In a parallel equation you have to deal with the slowest Link in that equation. This will determine how fast each layer in the network is going to go because your layer is only as fast as your slowest node. Serial equation simply refers to how fast a specific track of equations is able to calculate a result. Essentially, the serial equation will determine how fast you are able to go from one node to the next node to the output.

The best way that I could create a world example for this is a building of factory workers. Each level of factory workers can only hold so many Factory workers at a given time but each level that you add to it allows you to produce more products. Therefore, you have to find the rate at which the amount of factory workers and the amount of levels that you have produce a result that is successful at generating product at decent times. Therefore, linear testing is ensuring each worker is capable of working as fast you need them to. Parallel testing is ensuring how fast each floor of workers can produce results together. Finally, serial testing is how fast those results can be pushed out of the building as a final result.

Closed Testing

Closed testing is where your first bit of Randomness comes in. Sure, during the batches where you test for known values and then you also test for unknown values you do come across some Randomness but it's not true Randomness. During closed testing, you give the neural network access to variables that are now completely out of your control. Before, you could choose where the data was coming from and what the data pertaining to regardless of whether you knew of what the output would be or not.

Beta Testing

Beta testing is when you finally throw complete randomness at your neural network. During the initial testing you are just making sure that it works, and it works well. During closed testing you're creating redundancy at a small scale and during beta testing you're creating that redundancy on a grand scale. Beta testing is the hardest part of any program, regardless of whether it's a neural network or a software, because of the different case scenarios that the programmers simply could not account for because they have a very limited scope of possible scenarios.

Production

Production is the final stage of any particular program that's being developed at any given time. Production refers to the fact that the idea or concept is now within the products phase or the phase in which the company who invested in the venture is now able to make money off of it. It is during this time that the impact of the neural network can be measured over a given time span in the audience that it is designed for.

Expansion

Finally, if this was a hobby then you might not reach this stage but if it's a business than you ultimately want to reach this stage. The expansion stage is where you start the process all over again but instead of starting out with nothing but an idea, you are now finding areas where that idea can further increase its influence. The way that you find this out is you begin looking at the statistical variables gathered from the production cycle, which these will then tell you where your strong points are in the software and you can either make safe bets by finding a way to improve those points or you can take venture risks by investing in the weak spots so that you can flush those out.

Finally, We Come to A Close

Machine Learning with Python is Not Easy

Machine learning is a very hot subject right now and if it seems confusing at this point, it doesn't get any better. As much as one can try to explain it to another individual, machine learning is not a technique, it is not a school of math, it is a mathematical and programming concept that merges several different worlds of sciences together. You have the science of literature for text based neural networks, you have image recognition that takes in the photography and digital design sciences, and you have the predictive sciences that generally make up the rest of what machine learning is about. At first, it's going to seem like you're wading into a pool of mathematics that seem like they are everywhere at once because machine learning is a hot topic that a lot of people try to dumb down so that people get excited about it. The problem is that this doesn't do anyone any favors because machine learning is a complex mathematical art. It requires a significant understanding of statistics, calculus, programming, and conceptual analysis in order to effectively put it into place.

If you are a person who wanted to jump into machine learning because it sounded fun and cool, I completely understand. I see advertisements all the time about how one can go take a course and learn machine learning, but what they don't tell you in that courses that they only teach a very specific type of machine learning for a very specific type of application. Essentially, you have the people who understand what's going on taking advantage of the ignorance of the people who do not understand what's going on. This is why it can seem very confusing to navigate the academic and mathematical literature surrounding machine learning in the first place. That isn't to say that you can't create a neural network in the first year that you go about trying to learn how to perform machine learning, but it is exceedingly rare to create a commercially applicable program within the first year of attempting to learn the science of machine learning itself, especially if you don't feel that you don't have the mathematical capacity that such a science requires of its developer.

Machine Learning with Python is Easy

Having said that, you can easily learn how to program in Python very quickly and there are enough resources that a dedicated individual can learn how to program for machine learning rather quickly. Python has the necessary libraries in order to do this and there are enough companies behind this era of technology that you don't necessarily need the best of computers in order to do the work that you want to do. In fact, depending on how you want to go about doing this you don't actually need to learn how to program your own neural network because some of the companies have actually released API libraries that allow you to use an existing neural network without having to have the hardware.

That means that creating a neural network is easier than ever before and learning how to code a machine learning algorithm inside of python is also the easiest it has ever been. In this book, we have gone over all of the topics I've listed here in the conclusion and I've expanded a little bit further because this is such a vast topic that I want to give you as much information as I possibly can. Having said that, there is a lot more to learn about machine learning than the algorithms that I have listed here, how you can go about developing and deploying a machine learning algorithm, and how to find a profitable algorithm at that. The subject material may be intimidating but just keep in mind that the Python Community is there to back you when you are having trouble.

www.ingramcontent.com/pod-product-compliance
Lightning Source LLC
Chambersburg PA
CBHW071325210326
41597CB00015B/1359